Commercial and Inclusive Value Chains

Praise for the book

'An excellent addition to the literature on the integration of the poor into mainstream value chains, illustrating that even vulnerable households can contribute to economic growth and their own development.'
Linda Jones, international consultant, Canada

'There is a growing recognition that the principles of value chain management apply to all enterprises – not just major global corporations. The authors of this important new work demonstrate how these ideas can be utilized to create truly "inclusive" supply chains resulting in beneficial outcomes to all parties.'
Dr Martin Christopher, Emeritus Professor of Marketing and Logistics, Cranfield School of Management, Cranfield University, UK

'Importantly, the case studies in this book highlight not only the successes and benefits of "inclusive value chains" but also some of the challenges and potential pitfalls. This book will provide a useful reference for individuals and organizations involved in the planning and development of commercially viable, inclusive value chains in developing economies.'
Dr David Taylor, value chain analysis and improvement specialist and advisor, formerly Co-Director of the Food Process Innovation Unit, Cardiff University Business School, UK

Commercial and Inclusive Value Chains
Doing good and doing well

Edited by Malcolm Harper, John Belt and Rajeev Roy

Practical Action Publishing Ltd
The Schumacher Centre
Bourton on Dunsmore, Rugby,
Warwickshire CV23 9QZ, UK
www.practicalactionpublishing.org

© Malcolm Harper, John Belt and Rajeev Roy and the individual contributors, 2015

ISBN 978-1-85339-867-4 Hardback
ISBN 978-1-85339-868-1 Paperback
ISBN 978-1-78044-867-1 Library Ebook
ISBN 978-1-78044-868-8 Ebook

All rights reserved. No part of this publication may be reprinted or reproduced or utilized in any form or by any electronic, mechanical, or other means, now known or hereafter invented, including photocopying and recording, or in any information storage or retrieval system, without the written permission of the publishers.

A catalogue record for this book is available from the British Library.

The author has asserted his rights under the Copyright Designs and Patents Act 1988 to be identified as authors of this work.

Harper, M., Belt, J., and Roy, R., (eds) (2015) *Commercial and Inclusive Value Chains: doing good and doing well*, Rugby, UK: Practical Action Publishing <http://dx.doi.org/10.3362/9781780448671>

Since 1974, Practical Action Publishing has published and disseminated books and information in support of international development work throughout the world. Practical Action Publishing is a trading name of Practical Action Publishing Ltd (Company Reg. No. 1159018), the wholly owned publishing company of Practical Action. Practical Action Publishing trades only in support of its parent charity objectives and any profits are covenanted back to Practical Action (Charity Reg. No. 247257, Group VAT Registration No. 880 9924 76).

Cover photo: Sacks of rice stored at the Agricultural Business Centre, Kenema, Sierra Leone. These centres are supported by FAO. © FAO/Caroline Thomas
Cover design by Mercer Design
Indexed by Elizabeth Ball
Typeset by Allzone Digital Services Limited
Printed by Standartu Spaustive, http://www.stanadrd.lt,
    Vilnius, Lithuania

Contents

List of figures, tables and boxes	vii
About the editors	ix
Acknowledgements	x

1. Introduction – what this book tells us about commercial value chains that include the poor ... 1
   Malcolm Harper, John Belt and Rajeev Roy

Part One Non-food value chains ... 15

2. Khat from Ethiopia to Somaliland – a mild stimulant but a major income earner ... 17
   Abdirazak Warsame

3. Beer from bananas in Tanzania – a good drink and many good jobs ... 27
   Jimmy Ebong and Henri van der Land

4. Seed cotton production in South Rajasthan – preventing child labour ... 37
   Kulranjan Kujur and Vickram Kumar

5. Stove liners in Kenya – less pollution, less charcoal and more income ... 47
   Hugh Allen

6. Granite in Odisha – from Indian quarries to European kitchens, if government allows ... 59
   Malcolm Harper

7. Remittances – from the global diaspora to the poor in Somalia ... 71
   Abdi Abokor Yusuf

Part Two Commodity foods ... 79

8. Nyirefami millet – a traditional Tanzanian crop, marketed in a modern way ... 81
   Jimmy Ebong and Henri van der Land

9. Rice – smallholder farmers in Malawi can be profitably included ... 89
   Rollins Chitika

10. Angkor Rice – 50,000 Cambodian farmers growing for export ... 97
    Rajeev Roy

http://dx.doi.org/10.3362/9781780448671.000

11 Moksha Yug – Indian dairy farmers don't have to be in cooperatives *Chandrakanta Sahoo*	105
12 Suguna Poultry – decentralized village production is good business *Malcolm Harper, Rajeev Roy and Phanish Kumar*	115
Part Three Non-commodity foods	**125**
13 Green beans – from small farmers in Senegal to gourmets in Europe *Miet Maertens*	127
14 Odisha cashew nuts to global markets – value added all the way *Kulranjan Kujur*	143
15 Palm oil in Peru – small-scale farmers succeed where plantations failed *Rafael Meza*	151
16 Organic turmeric from eastern India – healthy spice and healthy earnings *Niraj Kumar*	161
17 Conclusions – what can we learn? *Malcolm Harper, John Belt and Rajeev Roy*	173

List of figures, tables and boxes

Figures

3.1	BIL's main sources of banana and its main end markets	29
3.2	Value chain map for BIL	34
4.1	Average sales price at different levels in the cotton seed value chain	40
5.1	The Kenya Ceramic Jiko	48
5.2	Jiko liner value chain map	54
5.3	Water filter value chain map	55
6.1	Main features of the granite value chain, per tonne in Indian rupees	63
7.1	Overview of a Dahabshiil money transfer process	73
8.1	Map of Tanzania showing Nyirefami's main sources of millet and end markets of millet flour	86
9.1	MHL and the rice value chain	91
10.1	The Angkor Rice value chain	100
11.1	Moksha Yug Access value chain	110
12.1	Business model of Suguna Poultry	117
13.1	Value of fresh fruit and vegetable exports from Senegal, 1995–2011 (US$ billion)	128
13.2	Bean export region in Senegal and studied communities	129
13.3	Diagram of the bean export supply chain from Senegal (situation in 2005)	131
13.4	Participation of local households as contracted suppliers and agro-industrial employees in the bean export chain, 1996–2005	135
13.5	Comparison of average farming household income from different sources	138
14.1	Cashew value chain in South Odisha	144
14.2	Selling price of cashew at different levels of the chain	147
15.1	Organization of the palm oil value chain its main participants	154
16.1	Operational value chain	166
16.2	The economics of the turmeric value chain	167
16.3	The price build-up in the turmeric value chain	170

Tables

3.1	Banana beer value chain	33
4.1	Cash surplus from a typical farmer's tenth of a hectare of cotton seed production (figures in US$)	41

5.1	Estimated income of Chujio Ceramics	53
5.2	Impact of Chuijo Jiko on household incomes	56
5.3	Impact of Chuijo Jiko on the environment	56
7.1	Value added along the chain for $100 remittance to Somaliland	73
8.1	Millet value chain: costs, revenues and margins at different levels in the chain	85
10.1	Results from the survey of smallholders contracted to Angkor Rice	101
10.2	Economics of a household producing Neang Malis for Angkor Rice	101
11.1	Development of milk production in India	106
11.2	Profit and loss account for a dairy cow	108
12.1	Value addition in the Suguna Poultry value chain	118
12.2	Economics of a 2500 chick poultry farm supplying to Suguna	121
12.3	PPI score before joining Suguna Poultry and at current levels	122
12.4	Characteristics of poultry farm employees	123
13.1	Comparison of household characteristics	137
14.1	Average cost of processing 100 kilos of cashew in a processing unit in Parlekhamundi	148
15.1	Production costs for 1 tonne of crude palm oil	156
17.1	The case study value chains in summary	174
17.2	Small poor players in value chains: 'SWOT' analysis from the point of view of value chain leaders	178

Boxes

3.1	Testimonies of two banana producers	35
8.1	Testimony of trader with Nyirefami Millet	87
10.1	Stories from two rice producers selling to Angkor Rice	103
11.1	Three dairy farmers supplying milk to MYA	111
12.1	Poultry farmer case study	123
16.1	Mr Naresh's story of turmeric and ginger cultivation	169

About the editors

Malcolm Harper is Emeritus Professor of Cranfield University in England, and has taught at the University of Nairobi and Cranfield University; he has worked since 1970 in enterprise development, microfinance and other approaches to poverty alleviation, mainly in India.

John Belt works for the Royal Tropical Institute (KIT) in Amsterdam. He is an agricultural economist with more than 20 years' field experience in agricultural development.

Rajeev Roy was an entrepreneur and currently teaches entrepreneurship. He has been engaged in several entrepreneurship development initiatives across the world. He is mentoring several start-ups in India.

Acknowledgements

This book is a collection of case studies that have been written by many different people, and each of the studies involved conversations with far larger numbers, including company managers and staff at all levels, and, most important, the small-scale producers and other members of the value chains on whose experiences the cases are based. We are grateful to all of them, and in particular to the poorer people who are benefitting from inclusion in the value chains but who are nevertheless still much poorer, and perhaps much busier, than any of those who will read the book. We often pester such people with questions about their assets and incomes, which we ourselves would be very reluctant to answer; we only hope that their forbearance will at least be indirectly rewarded by the improvement and increase in inclusive value chains that the book will inspire.

We must also thank the writers of the case studies. Their names and some brief information about them are included at the end of each study, but we were unable to include a number of other cases because the value chains they described were actually dependent on subsidies, or for other reasons; we are grateful to these writers as well.

Andrew Shepherd from CTA, the Technical Centre for Agricultural and Rural Cooperation, of Wageningen, supported the idea from the outset and provided valuable assistance and suggestions. The staff of KIT, the Royal Tropical Institute in Amsterdam, also provided support, and Gerit Holtland of Consultancy and Training for Rural Transformation, and Fair and Sustainable Ethiopia, gave us some useful ideas as to how development institutions should respond to the messages in the case studies. Thank you, to all of them.

Malcolm Harper, John Belt and Rajeev Roy

CHAPTER 1
Introduction – what this book tells us about commercial value chains that include the poor

Malcolm Harper, John Belt and Rajeev Roy

Abstract

This chapter discusses the theoretical foundations of the topic dealt with in the book: value chains in developing countries and whether they can be inclusive of poor people in ways that will materially benefit them. Value chains that have not received the attention of international development interventions are the focus of the study – to understand in what circumstances poor producers are seen as good business partners. Terms used frequently in the book are also defined to help the reader understand the cases and their contexts a little better. The difficulties of measuring impact of value chain involvement upon the producers are discussed, as well as the choice of the Progress out of Poverty Index to measure economic status. The cases include commodity products, non-commodity food products and other non-food items in a value chain. This chapter provides a short introduction to each case in the book.

Keywords: inclusive value chains; impact; Progress out of Poverty Index; small-scale producers; smallholder farmers

What this book is about

This book is about value chains. Not all value chains, but a particular class of such chains – 'inclusive' value chains – which include and substantially benefit large numbers of poor people. These people are often smallholder farmers, but they may also be artisans, or small-scale retailers or customers.

The 'development community', including United Nations (UN) agencies, bilateral donors such as the British Department for International Development (DFID) and the United States Agency for International Development (USAID), non-government organizations (NGOs) and the governments of poorer countries themselves, have in recent years become involved in promoting and assisting value chains, as it becomes recognized that economic development and the alleviation of poverty are unlikely to be achieved by the public sector alone; the private sector is seen as the main source of growth, and development assistance is increasingly a matter of partnership between public and private entities.

http://dx.doi.org/10.3362/9781780448671.001

For-profit businesses, which depend almost everywhere on trading goods and services with other businesses, also build value chains, not to alleviate poverty but because such chains are vital for their businesses. Some large businesses whose value chains happen to include and benefit poor people may present these value chains as part of their 'corporate social responsibility' (CSR) activities, but the value chains that are described in this book have not been developed by companies in order to achieve social goals, or to promote a favourable 'image', but because they are good business. The poor people from whom they buy their raw materials, or through or to whom they market their products, are their best partners from a commercial point of view. They can perform whatever functions are necessary, to higher quality standards, or more reliably, or less expensively, than any other suppliers, and it makes good business sense to work with them, and to pay them more than they could earn elsewhere, so that they will do their best to continue.

Business relationships between large corporations and small-scale farmers, or other poor people, are sometimes believed to be necessarily exploitative. It is argued that the main or only reason the corporations have chosen to work with them is because they are cheap, and also because they are plentiful; they can easily be replaced if they become dissatisfied or worn out. This certainly happens in many value chains, but we have tried to find examples that are profitable and 'sustainable' for all their members; it is worthwhile to pay the small-scale farmers or others more than the going rate for undifferentiated crops, or other supplies or services, because they can in some way provide better value for money.

What this book is for

The book is not intended in any way to make a definitive statement about what we have chosen to label 'inclusive commercial value chains'. We aim merely to show by several examples that it is possible and profitable for businesses to build and maintain such value chains, without subsidy or other non-commercial assistance.

We use the term 'value chain' because it briefly describes the sequence or network of businesses, farms, artisans, retail shops and other intermediaries that are needed to get any product to its final users. Many business people who are more concerned with profits than definitions use the terms 'supply chain' or 'distribution channel', but the word 'value' makes the point that the members of a chain do more than 'logistics', the transport and storage or 'place' functions. They add value in many different ways, including production and promotion, so that the final product or service is actually a combination of each of the 'four Ps', as famously conceptualized by McCarthy (1960): Product, Place and Promotion, which can finally be offered to the customer at a Price that represents good value.

We have ourselves been engaged in business in the past, not in global or multinational companies but in medium- and small-scale businesses. One of us was in the marine fishing industry in India for many years, one was a commercial consultant and another worked in international marketing of hardware from the United Kingdom. We had to develop relationships with numbers of suppliers and customers in order to deliver value to our final consumers and at the same time to maximize our own profits and to remunerate our other partners in such a way that they would continue to supply us and buy from us. We had to decide whether to buy direct from manufacturers, or to go through intermediary traders, and also whether to sell the completed products or services direct to the consumers or through a further succession of wholesalers, distributors, retailers or other 'middlemen' (or 'middle women').

We did not realize that we were developing 'value chains' or 'supply chains'; we were involved in marketing, in purchasing, in 'make or buy' decisions, in selecting and managing the optimum supply and marketing channels for our products. Porter (1985) showed how companies, and whole nations, could improve their competitive position by optimizing value chains as a whole, rather than seeing them only through the lens of the 'leader' of the chain. Raphael Kaplinsky (2000) went further, and showed how value chain analysis could be applied not only to maximize the profits of one firm within the chain but also to achieve a 'dynamic shifting of producer rents through the chain', thus improving the share of poor and disadvantaged producers or others.

What this book tells us

Prahalad (2005) writes of 'inclusive capitalism', where he discusses businesses that try to include the poor and underserved markets and consumers. While Prahalad's focus was primarily on the poor as a market, several others have specifically looked at possibilities of including the poor in productive processes (Fairbourne et al, 2007; Harper, 2010; Donovan and Stoian, 2012). These writers and others discuss how poor producers and other intermediary microenterprises such as village-based processors, itinerant vendors and others can be linked into 'modern' value chains, and businesses themselves are under increasing pressure to obtain external certification to show that those who are employed in their value chains are not being exploited.

Several donor agencies have promoted this type of activity by supporting value chains. Their 'value chain approach' supports the development of value chains where inputs, finance, institutional development, business development services or shared infrastructure may be subsidized. A number of multinational corporations have also promoted such value chains, supported in part by their CSR budgets, but these initiatives rarely become mainstream growth businesses.

We aim to demonstrate that including and benefitting the poor can do good and be good business, at the same time.

The managers of the businesses who developed and lead the value chains described in this book may or may not have value analysed the chains in order to maximize their own business's share of the profits, but it should be clear from the cases that they certainly did not make a conscious effort to maximize the shares of their partners in the chain, rich or poor, except insofar as this was necessary to ensure that they would be willing and able to play their part in the growth and sustainability of the total system. The value chain 'leaders' are socially responsible businesses, but not in order to say that they are; it is good business.

We do not suggest that donors, governments and others should never subsidize value chains in order to include and benefit smallholder farmers or other disadvantaged groups. Farmers everywhere, and particularly in the European Union and the United States, are often very heavily subsidized, often to the grave disadvantage of far less wealthy farmers who produce the same crops in poorer countries. Few governments of these poorer countries are rich enough, or misguided enough, to subsidize their farmers to the same level, but our case studies show that it is not always necessary to provide subsidy of any kind to enable the small producers or other poorer participants in a value chain to gain from their inclusion.

'Sustainability' is something of a watchword for many donors; this refers not to the permanence of the donor institutions themselves, although many of them appear to operate as if they are sure that their donations and subsidies will always be required, and that their objective to eliminate poverty will never be achieved. The intention is that the chain will be financially sustainable, in that all the participants will earn enough money from it so that they will wish to remain involved; environmentally sustainable, in that it will not destroy the natural environment on which it, and everyone else survive; and socially sustainable, in that it will benefit or at least not damage the societies that are affected by its operations.

When value chains are supported by donors they presumably intend that the chains will be sustainable, in that the subsidy will not have to be permanent. This does not always happen; such assistance often continues almost indefinitely, and during our search for cases we came across more than one value chain that had received substantial and not always productive subsidy long after it had been established.

Clearly a value chain that must always be subsidized is not 'sustainable', by any normal definition of the term. Some initial assistance may be needed to demonstrate that the value chain can pay its way, or cover its losses before it is large enough to be profitable, but if businesses invest in a value chain from scratch, without any subsidy, it is *prima facie* more likely to be a chain that will survive, and grow, because its fundamental economics are sound from the outset. We cannot predict whether the value chains that are described in these case studies will or will not prosper, but we suspect that they will last longer and grow larger than chains that depend on subsidy.

There are of course large numbers of such value chains everywhere, in the so-called 'developed' and 'developing countries', trading locally or internationally, satisfying their customers and also making a substantial contribution to the well-being of the low-income smallholders, artisans, traders or customers with whom they are partnered. It was, however, not easy to identify specific cases, or, once we had done so, to obtain the information that was necessary for a suitable case study.

Framework of the study

We used a multistage process to identify and choose the cases. First we publicized the opportunity through our own and other contacts; these were mainly in the 'development' community, rather than with 'real' business people. We asked interested contributors to send us a simple 'map' of the chain, telling us what the products were, approximately how many people were involved at each stage, and some basic data on the funding and economics of the chain. We did not at this stage reveal that we wanted only unsubsidized value chains, because we did not want prospective contributors to be tempted to conceal subsidies.

We received a large number of submissions; many were rejected because they had been subsidized, or they involved very small numbers, or the aspirant contributors were clearly anxious to promote rather than to describe the value chains. We also tried to include cases from a wide range of countries, which handled a number of different products. We short-listed apparently suitable cases, omitting many that had received donor support, and we also had to reject several cases at the initial draft stage, because it became clear that they had been 'aided'.

We were able, thanks to the generosity of KIT in Amsterdam and CTA in Wageningen, to offer a modest fee to the authors, but some companies were naturally reluctant to share confidential information about their own and their partners' costs and revenue, or to spare the necessary time. On one or two occasions, the managers whom we approached also expressed some irritation at the suggestion that they were 'doing good'; they had little respect for the numerous and often ineffective development programmes that they had observed in their own neighbourhoods, and had no wish to be associated with them.

We deliberately avoided the largest and best-known multinational corporations; some of their socially responsible value chains are already well known and have been heavily publicized, and it is sometimes difficult to distinguish between the business realities of the chain and the CSR that the chain leaders wish to publicize.

In spite of these difficulties, we were finally able to obtain a number of case studies that satisfied our criteria. We have tried to ensure that the 15 cases that we finally chose to include in the book are a reasonably representative sample. Seven are from Africa, of which two are about businesses in Somaliland. This may seem disproportionate in relation to that country's population, or even to that of all of Somalia. We felt, however, that the unusual status of both products, a semi-legal stimulant drug known as khat, and cash itself, and the fragile nature of Somaliland, where many international companies fear to do business, where there are few if any 'development projects' and where even foreign relief agencies' staff are often expected to travel with armed guards, was a reasonable guarantee that the value chains would be resilient. All we had to do was to ensure that those who participated in them were making a better living than before.

The remaining cases are from South East Asia and Latin America, and six are from India; this too may seem too many for one country, but it is important to note that the very poor population of India, one country, make up over a quarter of all the world's poor.

Much of the literature on value chains is focused on farm products produced by smallholders. Farming is the dominant source of livelihoods in most poorer countries, although their rural populations are rapidly being overtaken by city dwellers. Twelve of our fifteen cases involve farm products. They include 'traditional' crops such as rice, millet and cotton seed, but we have also included case studies on milk, fresh vegetables, poultry and palm oil, all of which demonstrate less familiar approaches to the production and marketing of such products. And, in addition to khat and cash, we have included case studies on polished granite slabs, cashew nuts and clay stove liners.

The cases do not follow a standard structure; their content is very different, and a standardized format would have made the collection tedious. We hope that each case has retained the individual style of its writer and its environment. We have attempted, however, to ensure that each case study includes certain basic information.

In addition to the obvious data about the country or region, the product or service that is produced by the value chain, and the approximate numbers of people who are involved, each case also includes a 'map' of the value chain, showing the links in the chain, making up the chain from the initial raw materials and other inputs through to the finished product or service, together with an estimate of how much value is added at each stage. This shows how the total price to the final customer is made up, and what proportion of this price is retained by each link.

We aimed to show how the value chains have 'significantly benefited' the poorer people who are involved; this requires information about their condition before they joined the chain, and at the time when our contributors carried out their studies. Incomes may seem to be the best measure of economic well-being, but it is notoriously difficult to measure anyone's income, and even some of our presumably literate and numerate readers might find it difficult to assess the extent to which they are better off today than they were say five years ago.

Fortunately, however, we were able to use the Grameen Foundation's Progress out of Poverty Index (PPI) (see www.progressoutofpoverty.org), which provides a remarkably simple and reasonably accurate measure of economic well-being, with tailor-made versions for all the countries we covered, with the exception of Somalia, for which the instrument of a neighbouring country was deemed to be appropriate. This five-minute questionnaire contains ten objective questions, selected and tested for each country, such as 'does the household currently own any cattle sheep or goats?', to which only a 'yes' or 'no' answer is required.

The answers to these questions have been shown to correlate with the likelihood of the respondent being above or below a relevant poverty line. Most of our contributors were able to administer this instrument to a small sample of the smallholders, producers, workers, traders or customers or other poorer people of whom large numbers were involved in the value chain. They asked their respondents to give the information as it was at present, and for a time just before they had joined the value chain.

It was not generally possible to ensure that the samples were representative, or of sufficient numbers to ensure statistical validity, and a few of our contributors preferred to use other data that had been collected earlier or were otherwise more conveniently available, but we hope that each case contains at least some evidence that the value chain it describes is not only inclusive but also that it substantially benefits some, if not all, of those whom it includes. We did not have the resources to obtain comparative data from control groups, in order to carry out 'randomized controlled trials', and we cannot thus make any definitive claims for attribution. The livelihoods of the chain participants who were interviewed may have improved for quite different reasons, or the respondents may have been atypical; this is *prima facie* unlikely, and we are reasonably confident that the people who were involved have benefitted to a similar extent to those who answered our questions.

We aim to show by a range of examples that private for-profit businesses that operate in poor countries can buy from and sell through large numbers of poor people, and can do this at prices that allow everyone to make a profit. Poor producers or others can significantly increase their incomes over whatever they were earning before, and larger businesses can at the same time make a profit out of the transactions, without having to be subsidized by donors, by government or from their own CSR budgets. It is good business for everyone.

We hope that the case studies will show not only that this is possible but how it can be done. Every case is of course different, and there are no standard solutions, but each of the 15 value chains that are described in the book illustrates how it has been possible for one product or service to be profitably produced and delivered in a way that involves and substantially benefits large numbers of poor people.

The cases

The 15 cases are divided into three sections. They relate to the products that are handled by the chains, not to the type of chain or to the lessons that can be drawn from it. Every reader will of course draw his or her own conclusions from each example, but there are certain specific points that emerge from each case, and from each section.

Many studies of value chains deal with internationally traded farm products. We have included some such cases in this collection, but the first section deals with unfamiliar products and services; we believe that these examples are less well known, and can thus perhaps teach us more than the more familiar commodities. We start with khat, a mild narcotic that is very popular with Somali people everywhere, but is illegal in the United States, the United Kingdom and some other countries. The khat leaves are highly perishable, and must be chewed within 24 hours of having been plucked. This case shows how a multilink chain makes it possible to satisfy this demand, across international borders, at a reasonable price, even in a region that has been subject to warfare and violence for the last 20 years. It is instructive, if sobering, to compare the effectiveness of this value chain with the experience of so many attempts to bring more desirable products to disadvantaged people living in remote areas.

The second case describes the value chain for banana beer in Tanzania. This low-cost local product, made with locally available materials, can successfully compete with global brands, and smallholders, traders and local entrepreneurs can benefit from the value chain; the dominance of international brands can be challenged.

Genetically modified cotton seed is another controversial product. Small-scale 'tribal' farmers in South Rajasthan in India have taken over this very labour-intensive crop from better-off farmers in neighbouring Gujarat. As a result, the employment of migrant children has more or less ceased; a complex value chain of organizers and agents links large numbers of tribal farmers to seed companies such as Monsanto, and the farmers' incomes are substantially increased.

Small free-standing charcoal stoves or *jikos* are widely used in East Africa. They are dangerous, they do not last and they are inefficient, and they produce noxious fumes and use large amounts of charcoal; this is costly for poor people to buy and for the dwindling tree cover. A local entrepreneur has built a business making clay liners for the stoves; he has focused on sound engineering and has also devised an ingenious decentralized distribution system for his products. The liners mitigate many of the problems, and the business has succeeded where many well-meaning but unbusiness-like donor-funded 'stove projects' have failed.

The next case is about the value chain for polished granite from India. It involves government, quarries, block cutting and polishing firms, and further distribution and processing businesses that finally offer high-quality durable surfaces for kitchens and other applications, worldwide. The industry can seriously damage the environment unless it is properly regulated, but the local government authorities have responded to this by forcing many of the quarries to close, thus destroying many jobs and depriving the economy of much-needed export earnings. The study of one company that has survived shows the potential that is being damaged by misguided official action.

The value added by each link in a value chain is usually measured in money, as it is in the tables that are included in each case study. The last case in this section, however, is about a value chain that moves money itself. Development agencies have recently 'discovered' cash transfers as a form of assistance for poor people, but remittances within and between countries have always been a vital lifeline that enables better-off relatives and others to send money direct to those whom they wish to help. These transfers have traditionally used informal *hawala* channels, but electronic funds transfer systems now enable these transactions to take place safely, at low cost and almost instantaneously, through complex chains of thousands of intermediaries, even to scattered and disadvantaged people in the Horn of Africa. This process has been dramatically simplified because of the widespread ownership of mobile phones, or of SIM cards that can be used in a neighbour's phone. Dahabshiil is privately owned and profitable, and has its own CSR programme, but it is paradoxical that its remittances and those of its competitors are now at risk of being stopped by the actions of a large multinational bank that claims to be a leader in universal banking in Africa.

The cases in the next section describe value chains that deal with more familiar commodities, whose promoters have devised innovative strategies that enable them and the farmers who supply them to make substantial profits and improvements to their livelihoods.

Millet is a traditional smallholder crop, grown mainly for subsistence, and its demand, like that for many such crops, has tended to decrease with 'development'. The Nyerfami company, however, has seen the opportunity in Tanzania for an upgraded product, properly processed, and has developed a multistage value chain with backward linkages to the farmers and forward linkages to consumers. The company provides millet farmers with threshing equipment that enables them to upgrade the product and make it suitable for further processing and onward distribution at much higher prices than are available for the undifferentiated traditional product. The company is profitable, and also enables its suppliers significantly to improve their incomes.

Malawi has ample land that is suitable for rice, but must nevertheless import large quantities of rice every year. Malawian farmers cultivate much less remunerative crops, for lack of knowledge, finance and suitable inputs. Mtalimanja Holdings, a local company, is starting to remedy this situation; they are helping smallholders to produce rice for them, rather than cultivating it themselves on a large scale. This requires less capital investment for the company, and is potentially very profitable. The company procures rice through local village traders and vendors, who make a reasonable profit for themselves, and the farmers can take advance loans and rent power tillers from the company. Many of the few hundred farmers who have thus far started supplying rice, mostly farming on less than one acre, have more than doubled their incomes. As the scheme expands, several thousand farmers are expected to enjoy similar benefits.

Cambodia is the world's eleventh largest producer of rice, and this includes substantial quantities of high quality aromatic rice, but its production per acre is less than half that of neighbouring Vietnam. A local entrepreneur realized in 1999 that this constituted an opportunity. He set up Angkor Rice to capitalize on the export demand, and decided that the best way to get local farmers to produce to the necessary quality standards was to work with their existing community organizations, not for political purposes but to help them to produce what his company needed. By 2013 some 50,000 farmers were under contract to supply the company. They use organic methods to ensure high quality, and high prices, but the company avoids the high costs of certification by strict quality control and effective branding. Some farmers are now exporting direct, bypassing Angkor Rice, but their production can be replaced by recruiting new farmers.

India is the world's largest producer of milk, thanks in large part to the success of the Amul cooperatives that have been promoted by the National Dairy Development Board; but Moksha Yug, based in the southern state of Karnataka, shows that a private company, without any farmer groups or associations, can profitably compete by reaching out to even smaller-scale producers than state-sponsored cooperatives whose members have to take their milk to village-based collection centres every day. Moksha Yug collects milk from 15,000 farmers, most of whom have only one animal. The collectors take the milk to the company's small-scale local chilling centres. They then sell it in bulk to other businesses, or through the company's own retail chain, which outsources the processing and packaging to other firms. Moksha Yug is growing strongly, and the main threat to its success is from political interests who resent the fact that a private company can compete with 'their' cooperatives.

Chicken meat is by far the most popular meat in India; it is cheaper than any other, and is not avoided for religious reasons, as is both beef and pork. Poultry can be grown on a very small scale, with as few as half a dozen birds, or in large factory-farming units holding several hundred thousand birds. The Suguna company, which is one of the top ten poultry producers worldwide, has chosen a middle route; they work with some 20,000 independent out-growers, who have an average of some 5,000 birds each. They are provided with chicks, feed and veterinary care and are then paid a standard price per kilo for the finished birds. Suguna collects them from the farmers and sells them through independent trade channels or through its own retail outlets.

The last section describes four value chains that handle less-familiar foodstuffs: green beans, cashews, palm oil and turmeric.

Fresh fruit and vegetable exports from Africa to Europe have increased dramatically in recent years, such that European consumers are no longer aware of the seasonality of many items; they are available year round. Green beans and other products have more or less replaced groundnuts, which used to be Senegal's main export crop. Fresh produce is subject to increasingly stringent regulations and controls, which require every batch to be traceable to the farm where it was grown. These requirements have lead to tighter relationships between members of the value chains, and exporters often find it easier to maintain the necessary standards when they cultivate their own crops, rather than buying from independent smallholders. As volumes have grown, some farmers have been able to conform to the rules, while many others have found employment with the large producers, often on under-used land that the companies lease from communities, sometimes on condition that they employ local people. Green bean cultivation is seasonal but very labour intensive; the workers on the large farms are relatively well paid, most of them are women, and their earnings are such that incomes of households that are involved in green bean production, whether on their own account or as wage earners, are double those in the same area that are not involved. Like most people in so called 'developed' countries, and indeed most readers of this book, most Senegalese vegetable cultivators are employed rather than self-employed.

Cashew nuts are a familiar and rather expensive snack food in Europe and North America; the value chains through which they reach 'Northern' consumers often require the raw material to be shipped from Africa to processing units in India, and thence to consumers in Europe and North America. Our case examines one such chain, in Odisha in India; in general, the village householders who cultivate and harvest the cashew trees benefit substantially from increasing international demand, as do some of the processing factory workers and those who are employed further down the chain. Most of the factory workers are women, and the process whereby the kernels are extracted from the fruit is delicate, laborious and painful, in that the fruit contains oil that can burn and damage workers' hands. Many campaigners have publicized the low wages and difficult conditions of Indian cashew nut workers, but there are few alternative employment opportunities in rural Odisha; the workers themselves are generally happy to have a job of any kind, rather than having to migrate to a nearby city or even to the other side of India.

Palm oil is usually cultivated on a very large scale, and smaller producers have in many places been forced out of the market by corporate plantations. The next case, from Peru, shows that a smaller, locally based value chain can sometimes succeed where larger-scale producers have failed. A local entrepreneur observed that a large acreage of oil palms had been abandoned. The scale was insufficient for high-grade processing, but he realized that local poultry and pig producers needed a lower cost replacement for the feed supplement for which they were using imported cotton seed oil. He succeeded in organizing the local people through their municipal associations, and they are substantially improving their incomes by harvesting oil palm fruits for his company. The fruits are transported to the processing unit by a network of transporters who are also increasing their earnings; after processing, a further network of transporters and other intermediaries brings the oil to its users.

The final case describes the value chain for turmeric, a rhizome similar to ginger, and an important constituent of curry powder, that is also used as a dye and a medicine. BioSourcing, a local company in Odisha state in India, has developed an effective and mutually profitable value chain for organic turmeric, which includes and benefits tribal farmers who are the traditional inhabitants of the forested hill areas of the state. The final product is sold to consumers in Europe, North America and Australia. Government forestry protection officers do not usually collaborate with for-profit businesses; they believe that they will always exploit tribal people and the forests where they live. BioSourcing, however, has successfully worked with government departments to promote groups of tribal farmers who cultivate and collect turmeric and at the same time protect the forests where the crop grows. The company keeps detailed records of every farmer, so that every batch of turmeric and the other items that they produce can be traced back to a particular farmer. The tribal people have never practised anything other than organic methods, and BioSourcing is able to exploit their traditional knowledge of their environment for the benefit of all parties. The company maintains an elaborate laboratory for continuous testing, and markets its products worldwide through the internet and through own-label packaging for a wide range of retail outlets, as well as selling through its own retail shop in Bhubaneswar, the state capital.

Every reader will draw his or her own conclusions as to what in particular has enabled each value chain to achieve this so-called 'double bottom line', to be both profitable and inclusive, but we conclude the book with our own views as to the reasons why the value chains are successfully 'doing well' for the private businesses that are involved, as well as 'doing good' to the poorer people who are also a part of each value chain.

References

Donovan, J. and Stoian, D. (2012) *5 Capitals: A Tool for Assessing the Poverty Impacts of Value Chain Development*, Technical Series, Technical Bulletin no. 55 CATIE, Costa Rica: Tropical Agricultural Research and Higher Education Center.

Fairbourne, J.S., Gibson, S.W. and Dyer, W.G. (2007) *MicroFranchising: Creating Wealth at the Bottom of the Pyramid*, Cheltenham: Edward Elgar Publishing.

Harper, M. (2010) *Inclusive Value Chains in India*, Singapore: World Scientific.

Kaplinsky, R. (2000) *Spreading the Gains from Globalisation: What Can be Learned from Value Chain Analysis?*, IDS working paper 110, Brighton: Institute of Development Studies.

McCarthy, Jerome E. (1960) *Basic Marketing. A Managerial Approach*, Homewood, IL: Irwin.

Porter, M.E. (1985) *Competitive Advantage: Creating and Sustaining Superior Performance*, New York: Simon and Schuster.

Prahalad, C.K. (2005) *The Fortune at the Bottom of the Pyramid: Eradicating Poverty Through Profits*, Philadelphia: Wharton School Publishing.

About the authors

Malcolm Harper is Emeritus Professor of Cranfield University in England, and has taught at the University of Nairobi and Cranfield University; he has worked since 1970 in enterprise development, microfinance and other approaches to poverty alleviation, mainly in India.

John Belt works for the Royal Tropical Institute (KIT) in Amsterdam. He is an agricultural economist with more than 20 years' field experience in agricultural development.

Rajeev Roy was an entrepreneur and currently teaches entrepreneurship in India and beyond. He has been engaged in several entrepreneurship development initiatives across the world. He is mentoring several start-ups in India.

PART ONE
Non-food value chains

CHAPTER 2
Khat from Ethiopia to Somaliland – a mild stimulant but a major income earner

Abdirazak Warsame

Abstract

This case study describes the value chain for khat, a mild narcotic stimulant, from Ethiopia to Somaliland. It covers the whole chain, starting in Ethiopia where it is grown, and ending with the consumers in Somaliland, including the actors in the chain and the income they earn. The information was obtained by the writer through field interviews at each stage, apart from in Ethiopia where it was necessary to rely on other informants. The study shows that the khat value chain benefits all its members, including the retail vendors in Somaliland who are generally poor women with no other sources of income. These benefits can compensate for some of the health and social disadvantages of the use of khat.

Keywords: Somaliland; Somalia; Ethiopia; khat; qat; women's microenterprises

Introduction

Khat is grown in the Yemen, Kenya and Ethiopia; the specific variety of khat described in this case study is grown in the Ethiopian highlands by a large number of different producers who have different sized farms.

The information in this case study was obtained from various actors in the value chain, through meetings, interviews and focus group discussions. This information was supplemented by personal observation of the different transactions in the chain, at the border and in Hargeisa, the capital of Somaliland. It was not possible to visit the farmers in Ethiopia. Many of the actors were reluctant to share financial information because of the highly competitive nature of the business, but many different actors, including some traders who had gone out of business because of heavy competition, provided valuable data.

The cultivation of khat

Khat is the leaf and twigs of the *catha edulis* tree, which is native to East Africa and the Yemen. This is an evergreen tree that can grow to 25 metres, with thin bark and many branches. It has to be liberally irrigated every day in order to achieve a steady harvest. Harvesting can start seven or eight years after planting, and the trees can continue producing for more than 50 years without replanting. Only young shoots are harvested, two to three times a week, in lengths of between 30 and 100 centimetres. There are several quality grades based on the origin, time of harvest, colour and tenderness of leaves. Pale leaves are considered to be of higher quality, but glowing leaves have a stronger stimulating effect. The estimated annual yield is 2 tonnes per hectare of fresh shoots from a well-managed *catha edulis* farm.

Khat must above all be fresh, and it is imported from Ethiopia to Somaliland every day. The producers are mainly small farmers. A small portion of the large quantity of khat that is used in Somaliland is grown locally on small farms located west of Hargeisa, the capital city, but this khat is considered to be of low quality. It has a lower price and consumers prefer the Ethiopian variety. Most of the khat consumed in Somaliland used to be grown locally, but many of the plantations were destroyed by the dictator Siad Barre in an effort to stamp out the habit. People have since then become accustomed to the higher quality that can be produced in Ethiopia because of higher rainfall. Some local khat is still consumed, however, and many small farmers grow it as a rain-fed cash crop and earn modest incomes from it, especially during and after the rainy seasons.

The origins of chewing khat

Although the history of the khat trade in Somaliland goes back to the late 1950s, consumption was limited at that time and was confined to a small number of devout followers of Sufism. These early consumers chewed khat in long sessions; they believed khat was a 'holy' plant that improved their ability to learn and to memorize. This belief arose from the fact that khat has mild psycho-stimulant effects, including increased energy, enhanced mood, reduced appetite and sleeplessness. These early khat users returned home to Somaliland after their religious studies in Harar in Ethiopia, and continued to use khat; the habit then spread to every place in the country, even to isolated villages. These 'holy men' can be regarded as the originators of the widespread use of khat in Somaliland today.

Transport from Ethiopia to Somaliland

Early every morning trucks and four-wheel-drive cars laden with khat drive from Ethiopia to the Somaliland border towns where they report to the customs authorities and pay the official import taxes. These vary depending on the quantity and quality of the product. There are no accurate figures on the daily quantity of khat entering Somaliland, because the border is porous and unknown quantities are imported illegally, but the official statistics from the customs authority suggest that over 300 tonnes (300,000 kilos) of khat of different varieties are imported from Ethiopia to Somaliland every day by companies and by individual traders.

Khat is highly perishable and must be very fresh when it reaches the final consumer in order to avoid loss in value to the user and loss of profit to the trader. Since freshness determines the price and the quality, the producers, traders, retailers and consumers pay maximum attention to ensuring that all the khat is of the highest quality. Rapid and reliable transport is critical, and the traders use well-maintained trucks and vans that can travel at high speeds and cover large distances in the shortest possible time. Khat has to be delivered on time and in good condition. The drivers must also be competent, experienced and careful.

Khat is picked in the late evening or very early morning in the Ethiopian highlands and has to be on sale in Hargeisa by midday or early afternoon. It takes about two hours to transport it from the farms to Hawaday in Ethiopia, where it is weighed and packed. This takes about four hours. The packaged khat is then taken to Jigjiga, which is a three-hour drive, and then two more hours to the border of Somaliland at Togwajaale. In another 90 minutes it is in Hargeisa, where the bundles of khat are thrown to the vendors from the trucks as they drive round the town at high speed.

Preserving the quality of the khat until it reaches the market is usually the responsibility of the trader or importer. Any loss in quality and freshness at any point in the supply chain leads to considerable reductions in prices or even to rejection of the product.

The importers, traders and vendors have accumulated a wealth of knowledge and market intelligence; they know how much khat of each type passes through the Somaliland customs, they are aware of potential market niches and market segments, and they know the up-to-date prices and who are their competitors. They get this information through unofficial networks of agents and other advisors.

How khat is used

The three main types of Ethiopian khat that are consumed in Somaliland are known locally as *dadar, jabis* and *fujis*. *Dadar* grows mostly in the Harar region of Ethiopia and was the original variety that was introduced to Somaliland in the late 1950s by the Sufi scholars. *Dadar* is reputed for its mildness and the low use of chemicals in its cultivation. It used to be the dominant variety but has more recently been displaced by *jabis*.

Jabis was introduced in the mid-1990s and is the most commonly used variety in Somaliland. *Jabis* has strong psycho-stimulant effects but it also contains residues of chemical fertilizers because it is intensively cultivated. *Fujis* came into the market in 2005 and like *jabis* is grown under intensive conditions. *Fujis* is the most expensive khat in Somaliland, and is only used by relatively affluent consumers.

Irrespective of the type consumed, the consumption of khat is widespread in Somaliland and is increasing rapidly, especially among young people and women. It is estimated that at least 20 per cent of the population are regular consumers. Khat will continue to be an important source of refreshment and recreation for many people, and the market is not likely to decrease in the near future.

There are some important economic issues that are associated with khat. It is an important source of income for poorer people, throughout the value chain. It may, however, be unhealthy for people who use it to excess. There is as yet no firm evidence for this, but the issue does merit research. Some working hours are lost because of khat use, in spite of businesses' efforts to impose regular working schedules. However, there are there are also some indirect economic benefits, as many khat traders invest their profits in other activities such as hotels, restaurants, filling stations, car washes, money-changers and grocery shops.

Advantages of khat for farmers

Smallholders face many problems, such as low prices, adverse weather, occurrence of pests and diseases and so on, and many progressive small-scale farmers move out of food production to cash crops that offer better market opportunities. For these farmers, khat has become very attractive and many of them have shifted to it.

Anecdotal evidence suggests that the number of farmers engaged in the commercial production of khat has increased substantially in recent times and that they have increased their incomes and improved their food security. Additionally, the Ethiopian federal government and some regional authorities earn considerable amounts of hard currency from taxes on the khat trade.

Khat is an important source of livelihood for its producers. It can be produced by smallholders on rain-fed land without irrigation, and it can also be grown more intensively with irrigation. Without irrigation, farmers can harvest khat twice a year, at the end of each rainy season. The average annual yield is between 700 and 1,000 kilos per hectare, and this is substantially more profitable than traditional crops such as millet or potatoes. Under the more intensive irrigated system, farmers can get two or three harvests every season, depending on the intensity of irrigation. These farmers can time their irrigation and synchronize their harvesting times with the times of highest demand and prices (TayeHailuFeyisa and Aune, 2003).The farm-gate prices of freshly harvested khat vary according to supply, type, quality and seasonality; the price of *jebis* for instance can range from 70 cents to US$2 per kilo.

Khat is usually handpicked in the early morning or late evening. As soon as it is harvested it is packed in sacks and loaded on to the buyers' trucks. The prices may have been agreed earlier by negotiation between the farmer and the trader, or they may negotiate at the farm gate itself. In such cases, the trader hires workers to pick the khat after the price has been agreed. The harvesting crew earn 30 cents a kilo.

Sometimes, the farmer and the trader hire brokers to negotiate on their behalf, for 10 cents a kilo. The traders may be directly involved in the initial purchase, transport, processing and exporting of khat, or they may participate only in the first two stages and hire someone else to do the processing. Processing requires special skills to sort, grade, weigh, pack and load the khat.

Apart from the price paid to the producers, the traders must also pay various taxes. A trader who is exporting to Hargeisa has to pay tax at the Somaliland border town of Wajaale, as well as transport and trade permits at each district along the route. The total varies but it is typically $1.30 per kilo; no other crop can generate such high tax revenues for government authorities.

Post-harvest operations

Jebis khat is normally packed into empty white sugar sacks. In order to protect the leaves from coming into direct contact with the sack, and to stop the fresh leaves from drying out, a bag of damp green grass is placed between the product and the outer sack. When the vendors get the khat they immediately put it into special wooden boxes that are stacked in the shade and covered with layers of wet sacks. This also reduces evaporation from the leaves. Finally, the vendors sell the khat to the final consumers in cheap plastic bags that are made in Somaliland.

When khat is shipped over long distances the drivers make several stops en route in order to open the sacks and cool the load with fresh air. The traders hire a caretaker who travels with the driver and supervises the whole process including the temperature and overall condition of the consignment.

If proper care is not taken, the quality deteriorates; the leaves darken or appear burned, and the khat starts to smell. Consumers refuse to buy it, or they demand a price reduction, which leads to losses throughout the value chain. *Dadar* khat can remain fresh for longer than *jabis*, but it is not so popular because it is not so strong.

Jebis is highly seasonal and supply depends on the rainfall in the highlands of Ethiopia, where it rains two to three times a year. The heavy rains that normally fall between April and October sometimes prevent the farmers from harvesting, but they usually manage to harvest what the traders have ordered. The high yields in the rainy seasons lead to excess supply and farmers have to compete to market their khat, but during the dry seasons, prices increase. This means that irrigation is a good investment. The prices paid to farmers range from 70 cents to $2 a kilo, although during winter they may go as high as $20 a kilo for short periods.

Consumers can usually determine the quality of khat by looking at it. If there are lots of small twigs and young leaves on the stalks it means good quality, whereas wilted old leaves indicate poor quality. The fresh young leaves, and occasionally the tender tips of the twigs, have the most gentle and stimulating effects.

Khat must usually be delivered to the market in the early afternoon. This is an enormous challenge, given the bad road conditions and other difficulties, but the trade is efficient and well organized.

The Hargeisa market

Hargeisa is the largest khat market in Somaliland. The first consignments arrive there around 9 to 10 am, and the larger shipments arrive between 2 and 3 pm, which is the time when most offices close. During the rainy season the roads become impassable and there are often delays, but the traders hire additional trucks to take over the loads from trucks that are stuck in the mud. The khat must get through. The process is described by Harper (2012) as follows:

> There was something medieval about the process. Nothing was mechanized. Everything was done by hand. Delicate weighing of the highest quality leaves – which fetch tens of dollars a bunch – was done with tiny brass weights on metal scales.
>
> In each and every room, a man sat with a giant ledger, carefully noting down weights, prices and other figures. 'This is the khat capital of the world', one man told me. 'This whole town is a khat factory. We harvest the leaves from the fields nearby, then we rush them to Awaday for sorting and pricing. This is our khat stock exchange. We all work from late afternoon until three or four in the morning. Only then do we rest'.
>
> In the middle of Awaday is a large five-storey building, the biggest I could see in the town. It is devoted entirely to khat and is divided into small rooms, where people prepare the leaves for transporting all over the world.
>
> The vehicles waiting on the road with their engines running reminded me of racehorses straining at the bit, before the starting bell. As soon as they are filled with khat, they race off into the night. 'This one's for Djibouti, this for Somaliland, Dire Dawa, Addis Ababa, London, China'.
>
> The reason for the speed is that khat has to be fresh. Wherever possible, it must be on the market before noon the next day. Otherwise, as one devoted chewer told me, 'it loses its deadliness'.

When a van arrives in Hargeisa the 'caretaker' who has accompanied it has quickly to count and deliver a predetermined quantity to each vendor. The quantity delivered to each vendor depends on his or her financial position and the number of customers he serves. Established vendors get between 50 and 20 kilos, and newly established ones receive only 5 to 10 kilos a day. There are about 250 vendors selling khat in Hargeisa and about 40 in Berbera.

Khat is taxed both in Ethiopia and in Somaliland. The Somaliland government collects SLSh9,000, or about $1.40, for each kilo of khat imported into Somaliland. Traders avoid this by declaring false weights to the customs officers, or by bribing them with khat, typically 2 or 3 kilos per load, which is equivalent to a 'tax' of 40 cents a kilo.

The khat vendors

The vendors pay a nominal of 10 cents daily municipal tax. Their kiosks are usually simple make-shift structures at inner-city locations such as main crossroads. These structures are unattractive and disorderly and cause congestion. The local authorities have more than once attempted to dismantle them, but without success. The mayor of Hargeisa is reported to have said that makeshift stalls would have to be demolished as part of his programme to revamp the city: 'We cannot accept small vendor-businesses in the street' (IRIN, 2012). But thus far the khat stalls are still in place, and it is unlikely that the authorities would destroy the livelihood of so many poor households in this way.

About 10 per cent of the khat vendors sell between 70 and 100 kilos a day; the majority, around 60 per cent, sell between 30 and 70 kilos a day, and the rest sell less than this. Many of the larger vendors employ one or two labourers to help them; they are paid between $5 and $10 a day.

The vendors have to maintain good relations with their regular customers; because khat is so perishable. They have to be sure to order enough to satisfy all their customers but if possible no more, since any unsold stock has to be sold off at a loss. The vendors manage quite large amounts of cash, and know when and to whom to allow credit. Some of them also act as money-changers, particularly by exchanging US dollars into Ethiopian Birr.

A vendor's gross profit is about $4 a kilo, and from this they have to pay any people they may employ at 50 cents a kilo. They also have to pay around 40 cents a kilo for electricity, water and polyethylene bags, and 30 cents for packaging.

Ifrah Ismail is a typical khat vendor. In 1995 she originally started selling *dadar*, the lower-quality type of khat, when she was a refugee in Ethiopia. When she returned to Hargeisa in 2005 she joined her sister Raho Ismail in the same business, but selling the higher-quality *jebis*. It was the only business she knew. The two sisters are both married and they have large families, but they earn enough to cover their daily living costs of around $15, and also to pay rent, school fees and electricity, as well as saving about 10 dollars every day.

It is hard work, they have to work long hours, and they sometimes have difficulty towards the end of the month in collecting debts from customers to whom they have sold on credit, but their earnings are reasonable and they are quite satisfied.

Hussein Omer started in the khat business in Hargeisa in 2006, after he finished school. At first he kept accounts for his father, who had been selling khat for many years, and when his father died he took over the whole business. He sells around 12 kilos of *jebis* quality khat every day, and makes about $45 profit.

He is pleased with his progress. He bought a taxi soon after he took over the khat business, which was also quite profitable, and he then sold it at a good price to cover his own marriage expenses, and to pay for the completion of his family's home. He also supports the education of his four young brothers and sisters.

He only has two significant problems. The traffic passes dangerously close to his stall, and some of his customers make him wait two or three days for his money; he sometimes has to take a loan from another business person to keep his own business going while he waits for his own customers to pay.

Many larger vendors employ teenage boys who help with sales, collecting cash from customers in the evenings and doing home delivery to important customers. Some larger vendors employ two assistants, who are usually paid a daily wage of 50 cents for every kilo their employer sells.

The 'caretakers' who are employed by the traders to safeguard the khat while it is being transported and until it is delivered to the final vendor, and who also stand in for the drivers when necessary, are paid a daily wage of 37 cents for every kilo that they supervise. These caretakers tend to move from one employer to another frequently.

The drivers do not usually own the trucks but are employed by the traders. They are paid around 5 cents for every kilo they transport, and they are also given 1 kilo of khat for their personal consumption.

The traders have their own distributors in the Somaliland markets. Their main job is to keep a record of how much khat is distributed to each vendor and to collect payment from them. They can suspend deliveries to vendors who default on their debts, and they are also authorized to sell off unsold khat at discounted prices at the end of the day.

Value-added at each stage

Poorer consumers wait for these discounted prices, particularly during the spring when there is high competition. The distributors earn 39 cents a kilo and they also get half a kilo of khat for themselves.

The traders have temporary storage and trans-shipment warehouses in the major towns in Somaliland and in Ethiopian towns such as Jigjiga and Wajaale. The warehouses are typically staffed by a foreman, an accountant and a transport supervisor, who each receive 30 cents a kilo. They may also employ a few labourers who are paid a minimal daily wage.

The traders pay the farmers anything between 70 cents and $2 a kilo, depending on the quality and the season, and it costs around $2.40 a kilo to transport it to the Somaliland border. The traders also have to pay the taxes and the cost of weighing and packaging. The prices vary according to the seasons and quality, but the traders typically sell the khat for $12 a kilo and make a margin of around $4. A trader typically handles between 100 and 700 kilos a day.

Transport from the border to Burao and Berbera costs around $2 a kilo, and because of the higher volumes and stiff competition, the cost to Hargeisa and Berbera can go as low as 10 cents a kilo. The round trip for a van load of 700 kilos from the border at Tog Wajaale to Hargeisa or Berbera uses 60 litres of diesel at $1 per litre. Some traders have their own vehicles and others hire contractors' vehicles.

Men are the main users of khat, although the number of women and young people who use it is rising steadily. Young people such as high school graduates even consume khat occasionally, mainly on Thursdays and Fridays.

Fujis is the most expensive type of khat and costs $40 a kilo. *Jebis* is the most popular type and costs $20 a kilo; the average consumer buys a 250 gram bundle of *jebis* at one time for about $5. The cheapest type is *dadar*, which is sold for $2 a bundle or even less.

The following list summarizes the value added at each stage for jebis, the most popular variety. These figures are necessarily very approximate because the prices vary according to the season and quality, but they give some indication of the large number of people who are involved and who earn their livelihood from khat:

- Farmer is paid by trader: $1.35 per kg
- Transport from farm to processing unit: $2.40 per kg
- Warehouse staff, three people paid 30 cents per kg: $0.90 per kg
- Transport from processing to Hargeisa: $2.00 per kg
- Driver incentive: $0.05 per kg
- Caretaker fee: $0.37 per kg
- Trader additional costs and margin: $5.03 per kg
- Total vendor buying cost from trader: $12.10 per kg
- Vendor costs and margin: $7.90 per kg
- Final retail selling price: $20.00 per kg

The plusses and minuses of khat

Most khat users say that they chew it because it is a pleasant social activity that they enjoy. According to Harper (2012):

> There is quite a ritual to chewing khat, which is usually done sprawled on the floor, preferably on a carpet or blanket, with cushions to lean on.
>
> Soft drinks, water and tea are placed before each chewer, together with a large bundle of khat, a bin for the stems and a cloth for wiping sweat from the brow. The session starts quite slowly. There is not much conversation as packing the leaves and stems into the mouth and chewing them are the priorities. After an hour or so, spirits lift, tongues fly and arms wave about. There is a lot of talking, planning, analysing, arguing and joking.

Chewing khat is mainly a leisure activity, but in Somaliland some people chew khat and work at the same time, usually after normal working hours.

About 300,000 kilos of khat are brought into Somaliland every day from Ethiopia. Around a third of this is consumed in Hargeisa, and slightly more than this is shared between the other two main cities of Berbera and Burao. The rest goes to smaller towns across the country.

Some types of khat are more perishable and cannot be transported over long distances. *Fujis* loses its freshness on long journeys and is therefore only used in places with good roads such as the main towns of Hargeisa, Berbera and Burao.

One other important by-product of the khat chain is the approximately 15 per cent of the total volume arriving in Somaliland that consists of hard leaves and twigs that cannot be chewed. This waste is discarded by the consumers and is used to feed livestock, mainly goats. This significantly reduces the feeding cost of livestock; goats in the main towns contribute 7 per cent of the total production of fresh milk in Somaliland.

In February 2014 the Somaliland minister of planning told the country's House of Elders that only 28 per cent of the adult men and 17 per cent of the women have jobs. The country's annual imports exceed its exports by about US$0.5 billion; although the deficit is more than covered by remittances from the diaspora, this lifeline cannot be expected to last indefinitely.

Khat has important disadvantages; it may have a negative impact on people's health, and it can reduce the productivity of those who chew it during working hours. It also creates large numbers of jobs, however, for men and for women; if more of it could be cultivated locally, it could make a major contribution to Somaliland's economy.

References

Feyisa, T.H. and Aune, J.B. (2003) 'Khat expansion in the Ethiopian Highlands: Effects on the farming system in Habro District', *Mountain Research and Development* 23(2): 185–189.

Harper, M. (2012) 'Frenetic pace of Ethiopia's khat boomtown', BBC, online at <http://www.bbc.co.uk/news/magazine-16756159> accessed 12 December 2014.

IRIN (2012) 'Hargeisa stalls demolitions infuriate traders', IRIN Humanitarian News and Analysis, 26 March, <http://www.irinnews.org/report/95162/somalia-hargeisa-stall-demolitions-infuriate-traders> accessed 23 June 2014.

About the author

Abdirazak Warsame is an independent consultant and works in agriculture, rural communications and with pastoral farmers in the Horn of Africa. He has worked for a number of UN agencies, NGOs and other development organizations.

CHAPTER 3

Beer from bananas in Tanzania – a good drink and many good jobs

Jimmy Ebong and Henri van der Land

Abstract

Banana Investment Limited (BIL) started as an informal, backyard business in 1989 and was registered in 1993. The company grew to become a market leader among local beer companies, controlling 50 per cent of the market share. On average the company produces 6,000 crates of banana beer in a day. Annual turnover of the company is US$5 million. This case study documents the inclusive value chain set up by BIL. Information was collected through key informant interviews held with various actors along the chain, focus group discussions with farmers and a survey among smallholder farmers supplying bananas to BIL. The value chain generates employment and income for smallholders, traders and factory workers.

Keywords: banana beer; Tanzania; banana value chain; local branding

The banana beer business

For a long time, banana has been one of the main food crops in Kilimanjaro and Arusha regions, particularly among the Chaga and Meru tribes. Since about a decade ago banana is no longer only a food crop but also a cash crop and currently banana plays an important role in the livelihoods of smallholder farmers. For many families the sales of bananas, usually generating an income all year round, provides quick cash needed to meet household needs. A large number of small-scale itinerant traders, many of them women, make a living from buying and selling bananas. Demand for both dessert bananas (also called sweet bananas) and cooking bananas is increasing, mainly from a growing population of urban consumers. Improved technologies for processing cooking bananas and a growing market for banana beer have given a push to banana production in Kilimanjaro and Arusha regions. For some smallholders, banana is now the main cash crop. The supply manager of BIL estimated that in Arusha and Kilimanjaro regions over 60 per cent of banana production goes to the brewing industry.

BIL was the first company to commercialize a local beer brew in Tanzania. After realizing the potential market for a low-priced beer for low-income consumers, Mr and Mrs Olomi established BIL as an informal, backyard

http://dx.doi.org/10.3362/9781780448671.003

business in 1989. They started by producing small volumes of banana beer, supplying nearby markets. BIL was registered in 1993 and since then the business has grown to become a formally structured medium-sized business.

BIL is not producing the cheapest banana beer in the market. The lowest priced beer is a simple homebrew, very much like BIL's first produce, low in nutritional content and often produced in unhygienic circumstances. Basically anyone can produce and sell this type of beer and it is therefore relatively easy to enter the market. BIL's beer however is more expensive, more nutritious and produced in a hygienic environment. Following the success of BIL, other companies are copying their business model and directly competing with BIL. Another category of beer is international and national brands, which are the most expensive, have a consistent taste, colour, quality and presentation, and are produced following standardized, certified, high-quality processing procedures.

Today BIL produces two main banana beers, sold as Raha Kamili and Raha Poa. Raha Kamili has a sweet taste, an opaque colour, pH of 3.5 and alcohol content of 10 per cent. Raha Poa is semi-sweet, with a golden opaque colour, pH of 3.5 and alcohol content of 11 per cent. Both are sold in 330 ml bottles. Both brands have a very high alcohol content compared to the top foreign and national brands like Heineken, Kilimanjaro, Castle and Serengeti. Raha Kamili and Raha Poa differ from local beer by their standardized properties (taste, colour, alcohol content, pH) and by being sold in bottles. The product is placed in between top-end brands and cheap local brews, specifically targeting low-income and middle-income consumers.

In 2005 the company was producing about 600 crates a day (15,000 bottles of 330 ml a day) supplying retailers in Arusha and Moshi. With a loan of $800,000 from the African Development Bank (ADB), BIL acquired a combination of Tanzanian, Indian and German technology and was able to automate a part of its processing line and increase its production. However, a large part of the processing is still done manually, providing employment to about 320 people. The investment increased production to the current average 6,000 crates (150,000 bottles) per day. The installed capacity can produce a maximum 60 million bottles per year. Currently BIL has an annual turnover of $5 million. It is now a leading producer and distributor of banana-based alcoholic beverages in the country and the market leader at the local level, controlling 50 per cent of the market. Its two beer brands, Raha Kamili and Raha Poa, are sold in 11 regions of Tanzania: Singida, Manyara, Tabora, Shinyanga, Arusha, Kilimanjaro, Tanga, Morogoro, Dodoma, Dar es Salaam and Coastal Regions.

Banana is sourced from various parts of the country. Currently BIL buys most of the 4,000 kilos of banana it requires daily from the Kilimanjaro and Arusha regions. In Arusha banana is mainly sourced from the areas around Mount Meru and Arusha. From October to December bananas are scarce in Arusha and Kilimanjaro so the company occasionally sources from Tanga and Morogoro and, once in a while, from Mbeya. On average 6,000 crates of BIL products are distributed to its main markets. Figure 3.1 shows the areas where BIL sources its banana and where it sells its beer.

Production and consumption of banana beer has a long history in Tanzania, especially among the Chaga people. Traditionally, banana beer,

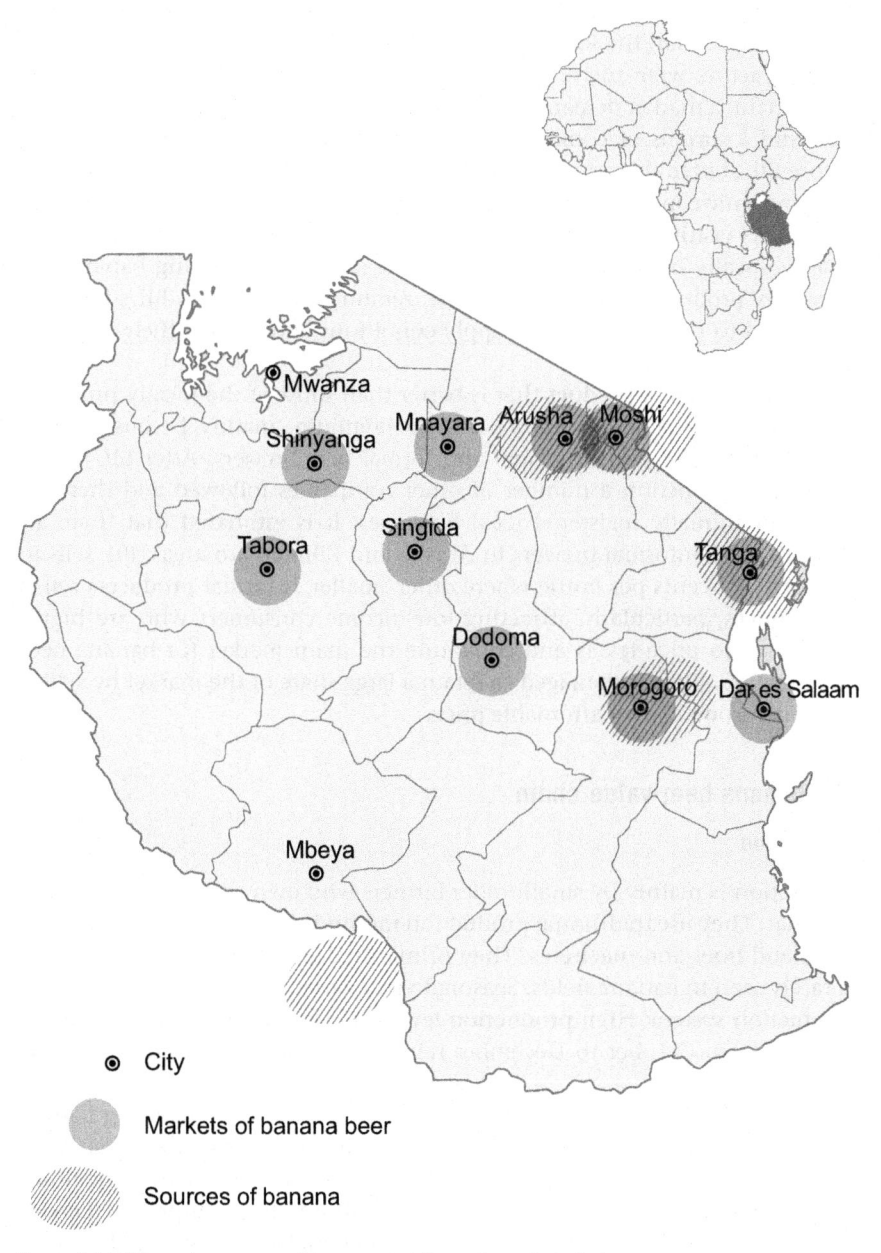

Figure 3.1 BIL's main sources of banana and its main end markets
Source: Based on information from BIL

known in the local language as *mbege*, was produced at a small scale and consumed at home. It was also one of the items used in Chaga customary marriages. Over the years, due to integration of cultures and customs, the consumption of banana beer spread and became popular among many Tanzanians. The nationwide spread of banana beer was also triggered by its being cheaper than the foreign and national beers.

Two factors were instrumental in the development of the banana beer value chain: a market demand for good quality, hygienically produced banana beer and a surplus of cooking bananas in Arusha and Kilimanjaro regions. Before BIL started its industrial production only informal producers, working in unhygienic conditions, were making banana beer. Low- and middle-income consumers visiting local bars were looking for an affordable, authentic, good quality, locally bottled banana beer. At the same time cooking banana was massively produced but lacked a stable, remunerative market. BIL's founders responded to the demand and supply conditions and created their own beer brand.

Despite having a product that is better than most of the locally produced beer, BIL is not able to capture a price premium, mainly because of stiff competition from both informal and formal beer brewers. After BIL started industrial production a number of other companies followed and there are now four formally registered local breweries. It is estimated that there are more than 40 informal brewers in Arusha and Kilimanjaro areas. BIL sells its beer at US27 cents per bottle where other smaller, informal producers sell at lower prices, particularly attracting low-income consumers who are highly responsive to price levels and constitute the main market for banana beer. Nevertheless, BIL has managed to retain a large share of the market by selling a quality product at an affordable price.

The banana beer value chain

Production

Production is mainly by smallholder farmers who own on average 2 acres of bananas. They use traditional production methods working with simple tools like hand hoes and machetes. They primarily depend on rainfall; irrigation is rarely seen in banana fields. Seasonal yield variation is a key feature of the production system. High production levels are from February to September, whereas from October to December rain is scarce and production volumes are low.

The most common means of propagating bananas is by planting banana suckers. Farmers obtain the suckers from their own farms. About ten years ago, after banana wilt affected the banana subsector, the Selian Agricultural Research Institute (SARI) produced tissue cultures from improved banana varieties. Some farmers obtained planting material from tissue culture and reproduced the improved variety on their farms.

Trade

BIL sources over 80 per cent of its banana from ten large-scale traders. Five of them are regular suppliers. Each of them sources from at least 50 smallholder farmers on a regular basis. Each trader has a designated day to supply banana to BIL. Payments are made two to three days after delivery. Sometimes BIL provides credit to the larger suppliers and in exceptional cases it can pay upon delivery.

There are around 20 farmers in the neighbourhood of the BIL factory who also act as traders and supply bananas to BIL at a small scale (each up to 10 kilos per transaction). In Meru there are a number of farmers who also act as traders, most of them women. They buy raw bananas from other small-scale farmers, and sometimes from other traders. They ripen, peel and sell the bananas to the larger traders supplying BIL. Generally bananas are delivered to BIL when they are fully ripe and peeled; this is done by the producers and sometimes by the traders.

About a year ago BIL was sourcing bananas from four farmer groups but it faced a number of challenges, mainly related to transportation and logistics. At the moment no farmer group is supplying bananas directly to BIL. Some farmers who previously were part of these groups are now selling their bananas individually to the large-scale traders.

Processing

BIL processes banana into beer by adding water, sugar, yeast and caramel. Fermenting and boiling is done manually, mainly by women labourers, but the brewing process is fully mechanized. BIL receives ripe and peeled banana from traders in special plastic bags. The bananas are weighed and boiled, and then banana juice is extracted and pumped into fermenting tanks, where yeast is added and it is left to ferment. The fermented banana juice is tested in the factory's laboratory to check if it has reached the right alcohol percentage and sugar content. The fermented banana juice from the different fermenting tanks is blended to make a uniform taste. The beer is then pumped into a bottling tank, from where it enters the bottling and labelling line. The bottles are packaged in crates and stored, ready for transportation to the market. BIL can produce a maximum of 60 million bottles banana beer per year. On average 6,000 crates (150,000 bottles) are produced daily.

The beer is bottled in Heineken non-returnable bottles. People collect used Heineken bottles and sell them to BIL. BIL also recycles these bottles; they are returned to the factory by the distributors supplying bars and retailers. Thirty-two mainly female casual workers are employed by BIL to remove labels and clean the bottles. Sometimes BIL buys new bottles from the two main producers of Heineken bottles in Tanzania, both based in Dar es Salaam.

Although the reuse of Heineken bottles made the business model of BIL more competitive, some competitors have started to use similar types of bottles, making it difficult for customers to differentiate competing products. Therefore, BIL intends to sell its beer in its own bottles, differently shaped from other beer bottles in the market. It has ordered 40,000 bottles from Kioo Limited, a South African company with a production facility in Dar es Salaam. Obviously this will increase BIL's production costs and reduce its margins. Each of these bottles costs BILTSh320 (about $2), whereas a used Heineken bottle costs TSh100, i.e. it is more than three times cheaper. Unlike the producers of the new Heineken bottles, the supplier requires upfront payment for the production of their own bottles.

Wholesale and distribution

BIL has two lines of distribution to deliver its products to the market: using its own trucks and working with distributors. It works with two distributors that also function as wholesalers. They pay upfront for the beer ($4.30 per crate) and are given targets regarding the volume of beer to sell within a given period of time. Most of the production (about 80 per cent) is delivered in Arusha, Moshi, Tanga, Morogoro, Dar es Salaam, Coast Regions and Dodoma by BIL's own trucks.

Retail and consumption

BIL targets middle- and low-income consumers who mainly buy their beer in bars and to a lesser extent in retail shops, in both urban and rural areas. The company fixes the price of its products at a maximum of 23 cents for a 330 ml bottle of Raha Kamili and 25 cents for a 330 ml bottle of Raha Poa.

Currently BIL does not pay duty on its banana beer products. A standard import duty is charged on imported beer brands like Heineken, and an excise duty of about 4 cents (TSh504) is charged on nationally produced beers like Castle and Kilimanjaro. Recently the Tanzania Revenue Authority (TRA) imposed on BIL a tax of about 3 cents (TSh 420) per litre of banana beer. BIL wrote a complaint letter but has not yet received a reply. BIL is protesting against excise duty on banana beer, pointing to the 40 local informal brewers who will not pay tax and will thereby gain a competitive advantage over BIL.

The BIL value chain

Table 3.1 presents the value chain for BIL and also the direct costs, revenues and gross margins along the value chain. Costs are calculated per kilo of bananas. The information on the cost of producing banana was obtained through a focus group discussion with a group of farmers in Meru. Only information on variable costs could be obtained.

Table 3.1 Banana beer value chain

	Farmer	Trader	BIL	Distributor	Retailer
Buying price	0.00	0.27	0.38	6.88	8.26
Direct cost ($/kg)	0.02	0.01	3.46	n.a.	n.a.
Total cost ($/kg)	0.02	0.28	3.84	n.a.	n.a.
Selling price ($/kg)	0.27	0.38	6.88	8.26	9.17
Margin ($/kg)	0.25	0.10	3.04	n.a.	n.a.
Gross margin (%)	93	26	44	n.a.	n.a.
Value added ($/ kg)	0.27	0.11	6.50	1.38	0.91
Share of end market price (%)	3	1	71	15	10

Source: Field data and information from BIL

At trade level, cost calculations were obtained from the large-scale traders since they were available and willing to provide the information. These traders are also the best (cheapest) option for BIL to source their bananas. Traders have established local contacts and live in the communities where they buy banana. They know the local dynamics of banana trade better than BIL. By far the most value added in the chain is generated by BIL, taking about 71 per cent of final market price. BIL's gross margin is 44 per cent, while for the farmers it is 91 per cent and the traders 26 per cent. Unfortunately, no data could be collected at the distributor and retail level.

Secondary actors in the banana beer value chain

Secondary actors providing support services to the banana beer value chain include Selian Agricultural Research Institute (SARI), Tanzania Chamber of Commerce, Industry and Agriculture (TCCIA), Tanzania Foods and Drugs Authority (TFDA), Tanzania Bureau of Standards (TBS), University of Dar es Salaam (UDSM) and Swedish International Development Cooperation Agency (SIDA).

SARI spearheaded research in banana and supported lead farmers by giving them access to tissue culture of improved bananas varieties. TCCIA helps their members in lobbying against unfair practices, including the new tax regime the government is planning to introduce. Mr Olomi, the managing director of BIL, is currently the chairman of the Arusha Branch of TCCIA. TFDA has certified both beer brands of BIL. Also BIL holds a license from TBS. Although TFDA and TBS do not inspect BIL regularly, the TFDA certification and TBS licence helps to increase consumer confidence and sales. BIL secured financial support from SIDA through UDSM to build a waste treatment plant.

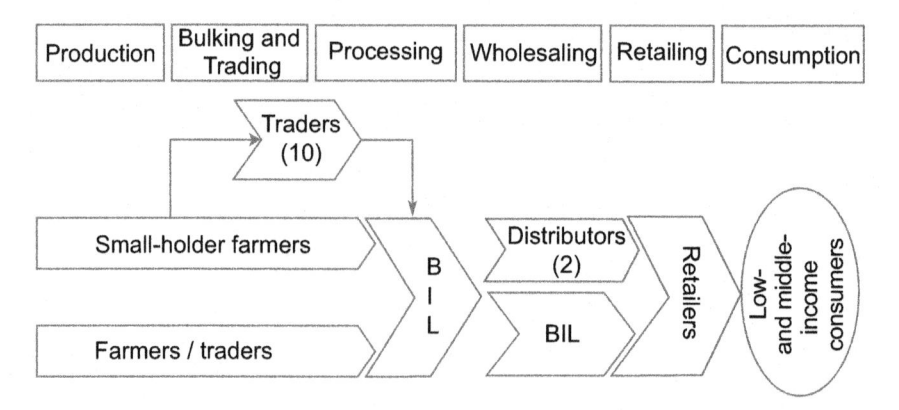

Figure 3.2 Value chain map for BIL

Why this value chain is inclusive

A survey was carried out among 20 smallholder farmers participating in the BIL value chain, using the Progress out of Poverty Index (PPI) methodology. From the collected information it appears that 5 per cent of the respondents considerably improved their living conditions thanks to their integration in the BIL banana value chain; they now earn more than $1.25 per day. It must be noted the number of respondents was small and sampling procedures were not scientific, and therefore no firm conclusions can be derived on the impact of the value chain on inclusiveness.

Besides the benefits for the smallholders supplying banana to BIL, a number of people with low incomes are directly or indirectly employed as a result of BIL business operations. BIL has 321 employees, 161 of them permanent (of which 37 are women), 63 semi-permanent (32 women) and about 97 casual staff (40 women). Casual and semi-permanent staff are predominantly from the low-income segment of the population and the number fluctuates between 150 during peak season and 40 during low season. Permanent staff earn a minimum of $90 and a maximum of $2,140 monthly. Casual and semi-permanent staff earn $4.50 per day. Other benefits for permanent staff include medical insurance for the staff members and their families, paid sick leave, meals at work and access to a credit facility through a Savings and Credit Cooperative Society (SACCOS). In addition to regular labour taxes such as Pay as You Earn (PAYE) and Skills Development Levy (SDL), BIL contributes to the national pension fund (National Social Security Fund or NSSF) for its semi-permanent staff. Casual labourers can access up to $200 as credit from Umbwe Development Association (UDEA), a credit facility set up for casual labourers. Repayment is deducted from their salaries. Access to credit has helped casual labourers to overcome financial problems, such as paying school fees or taking care of a sick family member.

> **Box 3.1 Testimonies from two banana producers**
>
> Mama Anoo is one of the women from Meru who both produces and trades bananas. She sells about 30 bananas every week to a BIL supplier and another 30 to other buyers. She is happy that BIL has created this opportunity for her by pioneering and professionalizing the banana beer business.
>
> Mr Maturo started growing bananas in 2003. He was one of the few farmers who was trained in banana production techniques and received 80 seedlings of banana tissue culture from Selian Agricultural Research Institute (SARI). He realized that bananas are a good option as they are a good source of food and income, can be fed to livestock and are a perfect intercrop for his main cash crop, coffee. He bought an additional acre of land in 2006 and expanded his farm to 5 acres; he also bought more livestock. Currently, banana contributes about 30 per cent of his monthly household income. With income from bananas, he has been able to educate four of his children; one of his children is graduating from the university this year. With banana, he is sure of an income ranging from $120–200 every week.

About the authors

Jimmy Ebong is a private sector development consultant who routinely conducts research and offers advisory services to businesses in Eastern and Southern Africa.

Henri van der Land is managing partner of Match Maker Group, a consultancy firm that conducts value chain studies, provides business development services and manages an investment fund targeting small businesses in East Africa.

CHAPTER 4

Seed cotton production in South Rajasthan – preventing child labour

Kulranjan Kujur and Vickram Kumar

Abstract

This case describes how cotton seed cultivation has shifted from the prosperous state of Gujarat in India to the poorer neighbouring state of Rajasthan. The farmers in Gujarat employed and often exploited child labourers who were forced to migrate from Rajasthan. Pressure from government and civil society made this more difficult, and the companies that previously procured seed from Gujarat have developed networks of farmer suppliers in Rajasthan itself. This process has involved the development of a multi-stage value chain, linking the major branded Bt cotton seed suppliers to the small tribal farmers. The result is that large numbers of small-scale tribal farmers have been able to increase their incomes significantly.

Keywords: child labour; Rajasthan; Gujarat; India; cotton; Bt cotton

Introduction

Shantilal is a poor tribal farmer supporting a family of seven members. He lives with them in Galandar village in the semi-arid region of South Rajasthan in western India, one of the poorest regions of the country. The Human Development Index of the region is lower than India's average, which is already in the lowest percentile in the world (UNDP, 2013). Shantilal wants to invest more in the education of his children as well as to improve his household's economic well-being.

In 2012 some of his dreams started to come true. He spent US$150 on education, which will help his children to complete a year of schooling. He also invested $1,300 in a concrete house, and saved around $500 to invest in agriculture for 2013 to ensure better returns. This was possible because he harvested Bt cotton worth $2,083 in 2012. This was ten times more than he earned eight years ago without cotton. In the last few years his income from cotton has increased from $400 to $2,083; this income can enable his household to move out of poverty.

Cotton has enormous potential to improve the livelihoods of poor tribal farmers in South Rajasthan. The area sown with cotton seed increases every year. According to the 2012 Statistical Abstract (Government of Rajasthan, 2012), it increased from 62,000 hectares in 2008–09 to 90,780 hectares in 2009–10, and anecdotal evidence suggests that it was more than 100,000 hectares in South Rajasthan alone. A large part of this is in the South Rajasthan districts of Dungarpur and Udaipur, where there are many farmers such as Shantilal. India has roughly a quarter of the world's cotton growing area and is the second largest global exporter. The nation has in the last ten years moved from being an importer to an cotton exporter, and cotton exports in 2009–10 were worth $1.67 billon. Imports in the same year were $192 million (Cotton Corporation of India Limited, 2011). This was mainly possible because Bt hybrid cotton seed was introduced in 2002, leading to spectacular growth in cotton production. In 2003–04 the area sown to Bt cotton hybrids was around 9 per cent of the total. This had increased to 66 per cent in 2006–07 (Venkateswarlu, 2007).

The leading cotton producing states in India are Maharashtra (27 per cent), Gujarat (18 per cent) and Andhra Pradesh (14 per cent). There is high demand, and many private companies have started buying cotton, so its cultivation has become very lucrative for farmers in Gujarat (Murugkar et al, 2007: 3781–89). Gujarat had roughly 9,600 hectares sown to cotton in 2006; in 2007 this had increased to 10,117 hectares and was increasing rapidly every year. Gujarat is relatively well off and farmers there hire labour from the neighbouring Rajasthan. South Rajasthan, with many poor tribal rural households, became an important source of cheap labour for cotton seed production in Gujarat. Cotton is a labour intensive crop. A hectare of corn, for instance, needs three people for a month, whereas one hectare of cotton seed production needs 25 labourers for two to three months. To save money, Gujarati growers started to employ children at lower wages; they claimed that children are better able to perform the work because their height is the same as the cotton plants. According to Venkateswarlu (2007), 150,000 children migrate from Rajasthan every year to work in the cotton fields of Gujarat. It is probable that some seed growers also traffic poor tribal children into work, accommodating them in inhumane conditions and even sexually harassing them. This continued despite the entrance of multinational companies in 2002, and finally local, national and international non-governmental organizations including UNICEF started to highlight the issue of child labour, which led both state governments to address the issue. Although there are no precise data, child trafficking has been reduced by action at its source and on the roads; people are aware of the issue and the industry is more cautious than before.

Cotton production also started to decrease in Gujarat; after growing at around 8.5 per cent a year from 2002–08, the area sown to cotton started to decrease. It dropped by 2.8 per cent in 2008–09, and cotton production decreased by 20 per cent from 2010–11 to 2012–13 (Cotton Corporation of India Limited, 2011). This decline can be attributed to the shortage of Bt cotton seed, which affected the whole country in 2008 and 2009.

Because of the shortage of labour and seed, growers in North Gujarat started to abandon seed cultivation. South Rajasthan became a good source of cotton seed and the government of Rajasthan's figures show that the acreage in cotton seed increased by almost 50 per cent between 2008–09 and 2009–10, and it continued to expand at 10 per cent annually thereafter. About 70 Gujarat-based entrepreneurs, known as organizers, are driving cotton seed production in South Rajasthan, and it is estimated that roughly 5,000 hectares are used for cotton seed in South Rajasthan alone, although the figure may be much higher. This dramatic shift to South Rajasthan has brought substantial benefits to the producers.

The cotton seed production value chain can boost poor tribal farmers' livelihoods. Although seed companies secure most of the value, it certainly benefits the farmers. In addition, other actors benefit, such as companies providing inputs, transporters, ginneries and other service providers in the chain.

Cotton seed production begins with the supply of cotton foundation seed by companies such as Monsanto, Mayhco and Nuvvizeedu who develop the seed. This is transferred to farmers through organizers who each commit to a target of at least $800,000 worth of seed and are responsible for ginning and processing the seed. The organizers are based in Gujarat and coordinate production with tribal farmers in South Rajasthan through local agents. They attempt to recruit agents who can work with as many farmers as possible. The farmers sow, cultivate and harvest the seed. The agents collect the harvested seed and take it to the nearest ginnery. The seed is ginned for the organizer, who supplies the good quality female seed to one of the major seed companies for final processing and branded packaging. The packaged and branded seed is sold to cotton growers through the normal farm input sales channels.

The organizers have established relationships with the big seed companies such as Monsanto, Bejoseethal, Vibha, Deepak Seeds, MICO, Bio Seed, Nanderi and others. The companies have formal agreements with the organizers, which are mostly family and farmer-run businesses in Gujarat. The organizers aggregate the seed on behalf of the seed companies and are the main investors of risk capital. The organizers in turn hire agents who are generally in direct contact with farmers for the supply of inputs and disbursement of credit.

The farmers are solely responsible for the production of the seed. After receiving foundation seed from the agents they start cultivation from late May to early June. It takes almost ten months from sowing to harvest. The harvested seed is collected and transported to the ginneries by the agents.

The ginneries separate the seed from the cotton fibre; the organizers pay for this, but the agents often cheat the farmers when calculating the final payments. The cotton fibre is sold through the regulated market in Gujarat from where it goes to spinners and weavers to be made into garments and so on for final consumers.

The seeds that have been separated at the ginneries are known as female seeds and are collected by the organizers. They are then sent back to the seed companies who sell the packaged and branded product to farmers at what appears to be a large profit margin.

40 COMMERCIAL AND INCLUSIVE VALUE CHAINS

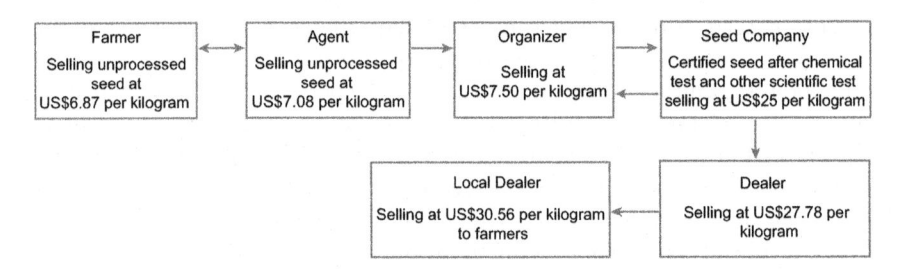

```
[Farmer]                    [Agent]                    [Organizer]              [Seed Company]
Selling unprocessed  →  Selling unprocessed  →  Selling at           ←  Certified seed after chemical
seed at                     seed at                     US$7.50 per kilogram     test and other scientific test
US$6.87 per kilogram        US$7.08 per kilogram                                 selling at US$25 per kilogram

                                                    [Local Dealer]                  [Dealer]
                                               Selling at US$30.56 per kilogram ←  Selling at US$27.78 per
                                                    to farmers                         kilogram
```

Figure 4.1 Average sales price at different levels in the cotton seed value chain

The value realized by the actors in the value chain is shown in Figure 4.1. The figures shown are the average selling price per kilogram for each actor in the value chain. Without including costs, nothing can be concluded on the profit made by the different actors.

The benefits accrued through the value chain

Although value is created for all the actors in the chain, this study is limited to understanding the benefits and roles of the two main actors in the context of South Rajasthan: the organizer and the farmer. This section discusses the type of farmers who are involved.

According to the Statistical Abstract (Government of Rajasthan, 2012), the tribal population of the two districts of Dungarpur and Banswara, which we call South Rajasthan in this study, make up almost 70 per cent of the total, with an average family size of five. Most of them live in rural areas. Although there has recently been some improvement in their children's education, under half of them go to school and many drop out. Most adults above 40 years of age are illiterate. The average yearly income of a tribal household is $521 (Murugkar et al, 2007) i.e. less than $2 per day. They have many sources of income, including agriculture, labour, forests and livestock. Agriculture is their main livelihood with a 40 per cent share, followed by casual labour with 30 per cent, and the balance is earned from 'minor forest products', such as collecting firewood, herbal plants and other products. Eighty per cent of the farmers are classified as small and marginal, their average land holding is under two hectares, and is usually far less. Most households only cultivate their land from June to October. The main crops are corn and pulses, which are mainly consumed by the farmers' households and are crucial for their food security. They also grow spices, vegetables and cotton as cash crops. On average they only use a tenth of a hectare for this.

The 3,270 farmers who produced cotton seed in 2012 in South Rajasthan were typical of these poor farmers. Each on average used 300 grams of foundation seed for his one-tenth hectare of land. The seed was provided to farmers by the organizer through the agent, at a total cost of about $7.50. Since the organizer had a buy-back arrangement with the farmers, the cost

Table 4.1 Cash surplus from a typical farmer's tenth of a hectare of cotton seed production (figures in US$)

	Cash outflow		Cash inflow	
Particulars	Amount	Particulars	Amount
Foundation seed	7.50	Ginned seed	465.80
Plant protection measures	59.60	Cotton	22.50
Cost of cultivation	67.10		
Interest on capital	13.40		
Cash outflow	80.50	Cash inflow	488.30
Cash surplus (inflow–outflow)	407.80		

Source: Authors' farmers survey, 2013

was borne by the organizer. This capital, which was effectively lent by the organizer, bore a 20 per cent interest charge for ten months. In addition to the seed, the cost of pesticides and insecticides was another $59.60. The total input cost on one tenth of a hectare of land was therefore $67.10 in 2012. The capital was usually borrowed from the organizer or from other informal sources at the same rate of 20 per cent for ten months. Thus the total cash outflow for ten months including interest was $80.50, as shown in Table 4.1. The labour for the cultivation was mostly provided by the household and this is excluded from the cash flow computation. The total credit requirement for the 3,270 farmers was slightly over $260,000.

The farmers sowed the crop in June 2012 and harvested it in February 2013. As shown in Table 4.1, the harvested production was aggregated and transported to the ginnery in Himmatnagar, where the seed and fibre were separated. The value of seed produced from a tenth of a hectare was $465.80, and the cotton fibre was worth $22.50. The total income was thus $488.30 and the agent then deducted $80.50, as calculated above. In some cases the agents also charged the cost of transport and ginning, which was between $5 and $7.50. The net cash surplus from one tenth of a hectare of cotton seed production was thus a little over $400, and the total cash surplus generated by the farmers in 2013, after deductions, was $1.33 million.

The impact of the value chain on the famers

Ideally, a value chain should transform the lives of the farmers who are part of it. The Progress Out of Poverty Index (PPI) questionnaire was administered to 22 farmers who were cultivating cotton seed. They were asked to recall the answers to the 10 PPI questions as they were before they started cultivating cotton seed, and for their situation in 2013. The average score obtained for the earlier period was 22.7, whereas for 2013 it was 30.65. According to the PPI measurement tool, this indicates that the likelihood of the households being under the poverty line of $1.25 per day was 49.7 per cent before they took up cotton seed cultivation, and in 2013 this was reduced to 30.5 per cent. This suggests that the households are moving out of poverty. This cannot be solely attributed to the cotton income, but a change of this magnitude can at least in part be attributed to what the farmers earned from the value chain.

An organizer – Patidar Agro Centre Company

The organizers play an active role in expanding the value chain in South Rajasthan. Although their aim is to maximize their own profits, they make a significant if perhaps unintended contribution to improve the livelihoods of the farmers with whom they work. There are more than seventy organizers operating in South Rajasthan. They worked with about 250,000 farmers in 2012, so that each organizer works with a little over 3,000 farmers. The total value of cotton seed production in 2012 was $111 million and is growing every year.

The Patidar Agro Centre Company is a typical organizer. The company is a registered business in Himmatnagar, Gujarat and has been operating since 2001. Its major activities are cotton seed production and marketing pesticides and insecticides. Until 2001 the company's owner was a cotton seed grower in Gujarat, who farmed 10 hectares. However, mainly because of a labour shortage, he was losing money. In order to maintain his income he saw that there was an opportunity to grow the crop in the neighbouring state of Rajasthan, which has a similar climate. In addition, he could minimize his own investment and production costs, because the farmers in Rajasthan would cover these costs. Also, there was little competition because very few others had seen the opportunity and no seed companies were directly involved in cotton seed production in South Rajasthan. The availability of skilled labour was also a factor as the rural population of South Rajasthan had experience of work in the cotton fields of Gujarat. Furthermore, he could earn more by selling inputs as there was high demand and limited supply in South Rajasthan. Seed production could be carried out without having a formal registered company, but not the sale of inputs. He therefore formed the Patidar Company in 2001 to procure cotton seed and supply inputs.

The company started cotton seed production in South Rajasthan in 2001, with fewer than 50 farmers who lived close to the state boundary. The business flourished due to the entry of multinational seed companies in Gujarat from 2002 and the Patidar Company aggressively expanded its reach in South Rajasthan. It became a one-stop shop for farmers taking up seed production. Along with foundation seed, it also sold inputs such as insecticides and pesticides. Importantly, the organizer provided inputs on credit, although only to a few farmers. In order to increase their earnings, and because the inputs were now available at one source, more growers in South Rajasthan started to participate in the value chain. In 2012, the company worked with 3,270 tribal cotton seed growers in the region, compared with the 50 with whom it had started in 2001.

The cotton seed business of the company has grown significantly. Precise figures between 2001 and 2007 are unavailable, but the company achieved average annual growth of 10.4 per cent from 2008 to 2012. The company's sales in 2008 were almost $900,000. This rose to $1.33 million in 2012, in spite of a small dip in 2010 due to a poor monsoon. Profit before tax was $20,000, or 2.2 per cent of the total turnover in 2008, and it increased to $43,333, or 3.3 per cent, in 2012. Despite regular profits, the management still feels the company is underperforming. With annual growth of 10 per cent it should be possible to generate a profit of at least 8 per cent of total sales. They feel that leakages at the agent level could be the reason. Much of the money supplied by the Patidar Company for disbursement as advance credit to farmers never reaches them. They discovered that the agents keep it. The agents deduct the advances from the final payments to farmers by charging exorbitant fees for processing. This deduction underestimates the gross production, eventually lowering profits for the farmers and for Patidar.

The agent – the link between the organizer and farmers

The agents are the critical link between the organizers and farmers. Five agents were interviewed for this study who had been associated with the organizer from the beginning of cotton seed farming in South Rajasthan. They are relatively better-off farmers and started cultivating cotton seed in 2001. The organizer covered their costs and provided technical support. They cultivated the cotton seed as instructed by the organizer, and in return the income from the production after deducting the initial investment was shared 60:40 between them and the organizer. This encouraged them to invest their own funds in cotton seed in subsequent years.

After assessing the farmers who wanted to be agents, including a review of their reputations within the community, the organizer invited them to become agents for scaling up cotton seed cultivation in the region. The organizer agreed to give them an incentive of $0.21 per kilogram, which was about 3 per cent more than the price they received as farmers. They accepted the offer and agreed to play their part in increasing the number of farmers cultivating cotton seed in South Rajasthan.

First, they met farmers in the region and convinced them to cultivate cotton seed. They advised the farmers and frequently visited their cotton fields. They also aggregated the farmers' needs for inputs such as foundation seed, insecticides and pesticides, and credit, so that the organizer could supply what was required. The agents then disbursed the money they received from the organizer to the farmers. After the farmers harvested the seed, the agents transported it to the ginneries. The cost of transportation and ginning was covered by the organizer.

The agents usually work in villages where they know the farmers, and on average each agent works with 80 farmers. Each farmer produced an average of 65 kilograms of cotton seed in 2012 and the gross income of each agent was almost $1,100. Assuming that each farmer produced 65 kilograms of seed on his plot, this means that the agents earned a total of well over $3 million in 2012.

The major cost incurred by the agents is travel to meet the farmers, the organizers and the ginneries. The actual cost of travel was difficult to calculate without records, but the agents said that approximately one third of the gross income from the work is spent on travel and other items such as tea or stationery. This would make the agent's net income about $7,230.

The agents hope that as the value chain develops they will be able to work with more farmers to increase their income. But they realize that reaching new farmers will be difficult because they only know farmers in their own immediate neighbourhood. They will be able to earn more only if they reach more farmers in the existing clusters. The organizers also prefer to appoint agents locally as it is easier for them to work with farmers whom they know.

The agents' major concern is that many farmers, who are interested in growing cotton seed, need credit. In addition to the investment required for cotton seed, they also ask for money to meet their other household needs. It is hard to provide all the credit they need, and farmers often borrow from other agents. To repay their debt they often want to sell their produce to them. The farmers repay the initial loans from the main agents, but when they sell to other agents this reduces the initial agents' earnings. About 5 per cent of the farmers do this every year.

Cotton seed can certainly improve the livelihoods of tribal farmers in South Rajasthan. Nevertheless, the value chain has only emerged in the last 15 years and it still needs to stabilize. Most of the ginneries are in Gujarat, which increases transport costs and thus reduces the farmers' margins. In time it should be profitable to start ginneries in Rajasthan. There are also other challenges that must be addressed if more farmers are to benefit.

Cotton seed production takes nearly 10 months from sowing to harvest, and farmers are often not paid for 11 months. This is a long period for tribal households who already have restricted cash flows. It is hard for them to have cash readily available all year round for their basic needs. This often discourages new farmers from taking up cotton seed production. The organizers can meet the capital needs of a few farmers, but rapid growth will require substantial additional credit. Assuming that each of the present 70 organizers works with about 3,000 farmers, they need $10.5 million for advances, and this demand is increasing by 10 per cent annually. Right now the farmers access funds from multiple sources, some of which are often exploitative. Their growing need for affordable credit must be satisfied.

In the present value chain, despite the organizer's best efforts, the farmers are highly dependent on the agents. Many of the agents are not transparent in their business transactions with the farmers or with the organizers. By keeping back 50 per cent of the money that is meant for credit, the agents are restricting expansion. The ginnery seed tests are also unclear, which may lead to unfair price deductions for the farmers, and the agents may also charge unfair rates for transport and ginning. This reduces the organizer's margins, and it further impoverishes the farmers. It is important, but difficult, to hire good agents.

The production of seed cotton is also more difficult technically than many other crops. Because of its long growing period, more care is needed in application of fertilizers and pesticides. Pest attacks are more likely than with other crops, which makes cotton seed more risky. Cotton is also a 'thirsty' crop, and the region is already short of water. All this means that the farmers need better ways of mitigating the risks. They rely heavily on their organizers through their agents. The organizer's capital is at stake, so they provide advisory services, but this becomes increasingly difficult as more farmers enter the value chain.

As the chain expands and includes more poor farmers, these challenges must be addressed with sustainable, economically sound and farmer-friendly interventions. It is hard to ensure that everyone in the value chain charges fair prices, but the situation in South Rajasthan is better than it was in Gujarat, which depended on child labour from Rajasthan. Although some farmers in Rajasthan have their children work in their cotton fields, they usually do this outside school hours. The cotton seed farmers say that their small plots only require family labour. It is very unlikely that outside labour, including children, will be employed because the cost will drastically lower the profit margin. The value chain should be closely monitored to see how it affects the lives of the poor and whether children are employed.

References

Cotton Corporation of India Limited (2001) 'Statistics', Cotton Corporation of India Limited, online at <http://cotcorp.gov.in/statistics.aspx?pageid=8> [last accessed 4.2.15]

Government of Rajasthan (2012) *Statistical Abstract 2012*, Directorate of Economics and Statistics, <http://statistics.rajasthan.gov.in/Files/Upload/S_ABST_VI_12-1.pdf> [last accessed 4.2.15]

Mathur, I. and Intodiya, S. (2010) 'Baseline Report of National Agriculture Innovation Project', Udaipur: Maharana Pratap University of Agriculture and Technology.

Murugkar, M., Ramaswami, B. and Selar, M. (2007) 'Competition and monopoly in the cottonseed market', *Economic and Political Weekly* 62(37), 15 September.

UNDP (United Nations Development Programme) (2013) *Human Development Report: The Rise of the South: Human Progress in a Diverse World*, New York: United Nations Development Programme.

Venkateswarlu, D. (2007) *Child Bondage continues in Indian Cotton Supply Chain*, Utrecht: India Committee of the Netherlands.

About the authors

Kulranjan Kujur is a Grants Portfolio Manager at the UK office of World Vision. He studied public affairs at Cornell University, USA. He is passionate about addressing poverty issues through the development and involvement of the private sector.

Vickram Kumar is a Programme Coordinator at ChildIndia Fund in Rajasthan. He has an MBA in rural management and is active in the promotion of sustainable rural enterprises. He has successfully promoted farmers' enterprises in Rajasthan.

CHAPTER 5

Stove liners in Kenya – less pollution, less charcoal and more income

Hugh Allen

Abstract

This case study of ceramic stove-liner making in Kenya illustrates the critical role of technology in creating competitive advantage and significant margins, while providing increased productivity and delivering improved product quality. It suggests that artisanal production cannot offer these things. It also illustrates that the transition from informal to formal status is incremental and best effected through steady reinvestment rather than debt, and that market access and development in the informal sector need creative thinking and a longer view, based on integrity and reliability.

Keywords: stove liners; Kenya; product quality; business profitability

Background

In 1985, at the age of 41, Kamwana Wambugu was at the peak of his profession. Headmaster of one of the best primary schools in Kiambu District in Kenya, he was a personable, successful and respected professional. However, as father to a family of six children, he felt the need to engage in business so that he could maintain the family's income beyond retirement.

With help from a friend who was a ceramic engineer, he decided to start manufacturing the insulating liner of the Kenya Ceramic Jiko (KCJ), which had been introduced into Kenya in the early 1980s by the Kenya Renewable Development Program (KREDP), based on a stove in widespread use in Thailand.

In comparison to the traditional charcoal stove, which it has mostly displaced, it is a revolution, saving the average urban family 350 kilos of charcoal annually: an estimated cost saving of US$14.50 a month, or $173 a year, or about 45 per cent of per capita income. This assumes that the user is buying by the bag. If, instead, fuel is bought daily, this increases the cost per kilo between 50 per cent and 100 per cent. Bearing in mind the weight conversion ratio of wood to charcoal of about 7:1, every stove offers annual fuel-wood savings of about 2.2 tonnes. Other advantages are its inherent safety because the flame is well insulated, and reduction in particulate emissions. In 2013 it was estimated that the KCJ is used in about 750,000 Kenyan homes, saving 1.65 million tonnes of biomass every year and $130 million in fuel costs (Global Alliance for Clean

http://dx.doi.org/10.3362/9781780448671.005

Figure 5.1 The Kenya Ceramic Jiko

Cookstoves, 2013). Wambugu played a critical role in making it successful by modernizing its manufacture and distribution. This is his story.

Start-up – following the herd

In 1984 there were several donor-funded initiatives to introduce the KCJ, most of them based on the idea of small-scale artisanal ceramics producers selling the liners to metal-smiths in the stove-making marketplaces of Nairobi, Kisumu and Nyeri. In 1985 this was the prevailing model and Wambugu became one of these small-scale liner-makers, using traditional manufacturing techniques.

After investing in materials, tools and working capital to pay a couple of potters, he was on the verge of shutting down just three months later having suffered heavy financial losses, equivalent to $6,000. The causes were easy to identify. Materials were of inconsistent quality; production techniques were of low productivity; losses in firing were unsustainable; product quality was variable and the stove liner often cracked on first use; and the workers Wambugu employed were unreliable, effectively holding the business to ransom.

Apart from financing the identification and adaptation of the stove to the Kenyan market, the promoting agencies did not focus on production technology nor product quality, relying instead on traditional hand-moulding techniques. Crucially they did not hire a ceramic engineer to solve the materials and production challenges and depended instead on employees with no ceramic engineering experience, whose private participation in liner manufacture was a clear conflict of interest.

It quickly became clear that technical leadership was lacking: by betting on informal producers, the promoting agencies had made it less likely that the right materials and production processes would be standardized. Soon, the KCJ acquired a reputation as prone to breakage and began to be less visible in the marketplace.

The evolution of technology – the source of competitive advantage

Wambugu was lucky. He had located a good source of clay along the Limuru–Naivasha road and, unlike nearly all other liner-makers, conducted experiments with a range of additive materials to improve its suitability. This resulted in a liner lasting between 2 and 2.5 years. This has been critical to building and maintaining a long-term competitive advantage. Nevertheless, the immediate challenges remained: how to improve productivity and make liners that were consistently sized, did not crack in drying and did not suffer firing losses.

The first issue that was tackled was moulding. Artisanal ceramic production for this product is inappropriate: no stove-maker wants to sift through dozens of liners to find one that will fit a casing, or have to customize casings to fit. It also created dependency on unreliable individual potters. Wambugu knew that something had to give and it looked probable that he would have to stop making liners.

The only solution was to mechanize the production process in order to increase productivity, reduce drying losses, ensure dimensional consistency and improve the finished appearance. Rather than having a skilled team of two moulders making 150 liners every week, he aimed to have a single unskilled operator manage at least 250 a day. If successful, this would leapfrog the competition.

Sketched out on scraps of paper, the design took shape and Wambugu used a Tanzanian friend's engineering workshop to make a moulding machine, known in the pottery industry as a jigger and jolley. This machine has a rigid arm and moulding blade descending into a spinning outer mould and shaping the liner. In the first eight-hour day it produced 310 liners. The issue of manufacturing productivity at its most critical bottleneck had been solved and labour costs cut by 95 per cent. Twenty-nine years later the same machine continues to function and has undergone no substantial rebuilding or redesign. A month after starting to use the machine Wambugu became the largest manufacturer in the country, producing and selling 800 liners every week and the threat of bankruptcy receded.

To reduce firing losses, he invested in a wood-fired updraught kiln, and to reduce drying losses he built a clay-mixing machine or pug-mill, which mixed the clay more thoroughly and in greater volumes. The first firing in 1985 was a success. Only 13 liners out of 443 cracked, while fuel consumption was reduced by two-thirds. Once this equipment was in place, there was not much technological change for several years, except that more jigger and jolley moulding machines were added, plus an industrial quality hammer mill. The pug-mill and kiln have now been superseded by equipment of much larger capacity.

Nothing that Wambugu did was revolutionary – except that no one else followed him down the same road, especially with respect to materials engineering and investment in mass production, using high-quality locally manufactured electrically powered machinery. Many visitors from competing factories observed what he was doing, but nobody copied him.

If there is an important lesson from Wambugu's experience it is that a competitive position is derived not only from efficiency and attention to detail, but the creation of barriers to competition, based mostly on investment in technology that allows for attractive returns and creates a safe environment for long-term fixed-asset investment, time for attention to market development and a more stratified organizational structure.

This is a basic lesson, but is often ignored by the legions of donor-funded appropriate technology advisors who champion decentralized, low-technology production, forgetting that quality and reliability in the market nearly always calls for some degree of technical specialization, centralized mass production and distribution networks.

Marketing and the supply chain

Up to now we have focused on how the business climbed out of near-bankruptcy by using solid technology, but in many ways, this is the less interesting part of the story.

In 1986 and still today, stove-makers in Nairobi are clustered in Shauri Moyo, where maybe a couple of thousand metal-smiths produce everything from wheelbarrows, to cookware, stoves and farm implements. One of the biggest products has always been charcoal stoves and the main early effort of the promoting agencies was to create linkages between these stove-makers and liner producers. Shauri Mayo therefore became the rich plum for liner-makers and in very short order intense competition emerged between the three or four main producers, with discounting, dumping of old stock and credit to stove-makers (a fatal proposition) being among the most popular (and counterproductive) tactics.

Wambugu was a latecomer to this market but decided, as a matter of business principle, to establish a reputation for excellent product quality, stable pricing and reliable delivery – and to avoid offering credit (with which he had a negative early experience). If he said he would make a delivery at midday on Thursday, he made sure he was there, and stove liner prices only rose incrementally. The price of a standard liner in 1985 was KSh15 ($1.25). Today it is KSh35 ($0.40), which is actually 60 per cent lower when corrected for inflation.

Over the long run these principles have stood him in good stead, but initially they cut no ice. He realized that everyone in Shauri Moyo was fighting over a bone that wasn't getting bigger. He began to look outside Shauri Moyo, at groups of two to three metal-smiths in local pockets all over the town, each of whom had to travel daily to Shauri Moyo to get their stock of liners. He began to visit these stove-makers, at agreed times throughout the week. This in turn attracted other stove-makers to these clusters and, by degrees, the satellite markets aggregated to a scale that is now much bigger than Shauri Moyo.

While he was developing this network, the costs of running around all of these points of sale on a daily basis and selling 10 liners here, 20 there, was too high, since his factory is 35 kilometres from the centre of Nairobi, in the uplands of Tigoni. He decided that he needed to build stocks of liners on site, so he negotiated with a single metal-smith in each of these locations to rent a small 1 cubic metre storage space, in which he built a lockable weldmesh cage. This enabled him to rationalize and reduce his supply trips, but still meant that he had to travel to each location daily.

To eliminate this constraint, he hired a relative who commuted along a daily circuit, opening up each weldmesh cage at fixed times of the day for about an hour. This met the stove-makers' need for daily purchase of a small numbers of liners at low cost, and covered an extensive geographical area, with over 20 such outlets servicing between 100 and 150 stove-makers.

The strategy of low-cost decentralized marketing is based on simple principles: always maintain a large stock of liners to allow for choice; always show up and deliver when you say you will; maintain consistent pricing, without spikes or deep discounts; and do not give credit.

Early lessons learned

Many times in the early years Wambugu applied to banks and non-government organizations for credit to expand, but he did not fit the approved profile of an entrepreneur. There was little curiosity from those who interviewed him as they ticked their boxes, asked all the wrong questions, did not visit his business, failed to recognize his special commitment to quality, technology and creative marketing and failed to get mud on their boots. In retrospect, not getting a loan was the best thing that could have happened to him. A loan would have enabled him to expand rapidly, possibly beyond his ability to manage the changes it would have demanded. Instead, both technical and commercial growth has been incremental, based on reinvestment and experimentation that has evolved into standard practice – and his anxiety has been much lower than if he had a large debt hanging over his head.

Although the decision to invest in modern technology was expensive, he quickly saw how it created a competitive advantage, and he continues to upgrade 29 years later. The design principles for all of the machinery were to allow a much larger capacity than immediately required, ensure that all machinery was of a high quality and durable, easy to maintain, powered by electricity rather than manually, and locally manufactured but by people who knew what they were doing, not in informal workshops.

Instinctively, Wambugu understood the four Ps of marketing – product, price, place and promotion – to build a value chain based on customer service and low-cost maintenance. He had the right mixture of products and built a large capacity to maintain consistent supply. His price was competitive because his investment in technology delivered a decent gross margin, even in the face of intense competition. He stuck to a stable price and avoided offering credit to a clientele unlikely to repay. He developed low-cost delivery channels and built satellite inventory at the point of sale, covering a much larger geographic area than his competitors and in aggregate, serving a much larger number of stove-makers on site. Promotion was based on direct marketing, with no advertising. By putting the product on display at the point of sale, no special promotion was needed.

A mantra of small business development is that small businesses need to keep formal records. This is nonsense. They need to be aware of what they owe, what they are owed and to be able to predict income and expenditure over a one to three month time horizon, or less, if they are selling vegetables in the local market. This can, of course, be all written down, or it can be kept in the head. In a family enterprise, where cash is king, the need for a conventional set of records is less important than being on top of combined family and business cash flow, and this was the guiding principle of Wambugu's business for its first 25 years, during which time it went from a backyard operation to an employer of 43 people. It is only since increasing profitability through the introduction of a water filter and engaging with the formal sector that the business has become more formal and has adopted a more conventional system of financial management.

The Great Leap Forward (but with better outcomes)

After the start-up and growth years between 1985 and 1990, Chujio, which was now the name of the company, ran at a steady state, making much the same range of stove-liner products, although adding a line of flower pots to its product mix. Expansion was incremental, based on reinvestment. During this time Wambugu and his wife Agatha put five of their six children through university, bought the land from which they mine the clay, built a three-bedroom house and acquired land on the Thika Road for a new factory that will be built in the next few years. The business met their needs for income and allowed them to save and diversify their investments.

At this point the story could have ended: an interesting tale of how an informal business used technology to create competitive advantage to survive and to satisfy a family's needs to educate its children and provide for old age, but serious growth needed new products with large gross margins. This called for high value addition in products that would be hard to copy technically (but needed to be compatible with the technology already in use) and could command a much higher price in a seller's market.

The breakthrough came in 2007, when Chujio was approached by Potters for Peace (PfP), who were looking for a manufacturer of a ceramic water filter. This product called for much higher technical standards than the Jiko liner, but was based on similar technology. PfP sent two technicians to train Chujio staff and the technology was quickly mastered. Once again, however, Wambugu sought to improve the quality and productivity of the manufacturing equipment. By now he had shifted from being a user and manager of technology into a creative practical engineer. He bought discarded Caterpillar hydraulic rams and linked them to hydraulic motors, which increased press productivity from 5 filters per hour to about 40. He built four of these presses and later supplied them to producers in other countries where PfP also worked.

He then turned his attention to the firing process and adapted one of his large updraught kilns to accept the filter. These investments greatly improved product quality and output, while reducing costs. Gross margin increased to about 50 per cent, and the price of the assembled product was 45 times more than the Jiko liner, while using less labour. This has been decisive in taking the business to a whole new level.

Precise income figures are not available; the owners are understandably unwilling to share confidential business data. The figures that follow are based on production figures that are available, when prices are factored in. Current liner production can reach 1,000 units a day, but a weekly figure, averaged over the year would be about 70 per cent of that amount: we estimate that net daily sales are probably about 700 units.

As the figures in Table 5.1 illustrate, the water filter is a game-changer, contributing about 80 per cent of total current income and as much as 85 per cent of gross margin. It is the key contributor in transforming the business from a large-scale informal operation, always at risk of external exploitation, into a successful and highly profitable quasi-formal enterprise whose future will demand continuing investment in research and careful selection of new products and technology.

Table 5.1 Estimated income of Chujio Ceramics

Item	Annual sales (US$)	Estimated gross margin (%)	Estimated gross margin (US$)
Liners	76,500	40	30,600
Flower pots	22,000	30	6,600
Water filters	425,000	50	212,500
Total	532,500		249,700

Value chains

Chujio Ceramics has always sought to be as vertically integrated as possible, so as to be in control of input supply, production and marketing, but the KCJ liner is itself an input into a stove that is finished by stove-makers. The water filter, by contrast, is produced as a multi-component, professionally packaged finished product, sold mainly to institutional buyers, with few retail sales.

Figure 5.2 presents the KJC stove liner value chain and Figure 5.3 the water filter value chain. KCJs and water filters.

Figure 5.2 shows a moderate degree of vertical integration. The acquisition of a 7-tonne truck means that all haulage activities are internal to Chujio,

54   COMMERCIAL AND INCLUSIVE VALUE CHAINS

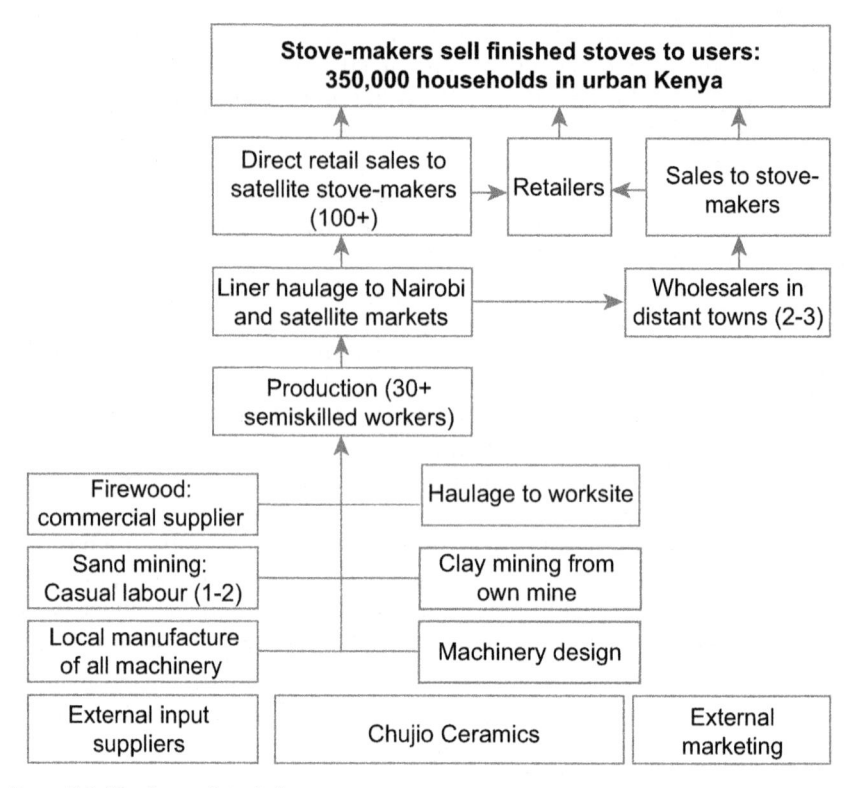

```
                ┌─────────────────────────────────────────┐
                │  Stove-makers sell finished stoves to users: │
                │      350,000 households in urban Kenya       │
                └─────────────────────────────────────────┘
                       ▲           ▲            ▲
        ┌──────────────────────┐  ┌─────────┐  ┌──────────────┐
        │ Direct retail sales to│  │Retailers│◄─│Sales to stove-│
        │ satellite stove-makers│─▶│         │  │   makers      │
        │        (100+)         │  └─────────┘  └──────────────┘
        └──────────────────────┘                      ▲
                       ▲                              │
        ┌──────────────────────┐            ┌──────────────────┐
        │ Liner haulage to Nairobi│─────────▶│  Wholesalers in  │
        │  and satellite markets  │          │distant towns (2-3)│
        └──────────────────────┘            └──────────────────┘
                       ▲
        ┌──────────────────────┐
        │    Production (30+    │
        │  semiskilled workers) │
        └──────────────────────┘
                       ▲
  ┌─────────────────┐      ┌─────────────────────┐
  │   Firewood:     │      │ Haulage to worksite │
  │commercial supplier│    └─────────────────────┘
  └─────────────────┘
  ┌─────────────────┐      ┌─────────────────────┐
  │  Sand mining:   │      │  Clay mining from   │
  │Casual labour (1-2)│    │      own mine       │
  └─────────────────┘      └─────────────────────┘
  ┌─────────────────┐      ┌─────────────────────┐
  │Local manufacture│      │  Machinery design   │
  │  of all machinery│     └─────────────────────┘
  └─────────────────┘
  ┌─────────────────┐  ┌──────────────────┐  ┌─────────────┐
  │ External input  │  │ Chujio Ceramics  │  │  External   │
  │    suppliers    │  │                  │  │  marketing  │
  └─────────────────┘  └──────────────────┘  └─────────────┘
```

Figure 5.2 Jiko liner value chain map

which maintains control over retailing to stove-makers – but not to end users. The only inputs that are supplied commercially, sand and firewood, are available from a local, low-cost market. The only part of the production process that Chujio does not control is that of final assembly into a completed stove.

Wambugu has used his technological advantage to gain a foothold into a competitive market, but has secured it by building relationships in the value chain that are more extensive and personal than those of his competition. By focusing on satellite markets and creating an efficient low-cost marketing infrastructure, he has built a system that is all the stronger for his investment in the personal relationships that are at its core.

In contrast, the value chain map for the water filter in Figure 5.3, while showing a higher degree of vertical integration than the Jiko liner, also shows a dependency on a very limited and undependable cadre of institutional buyers, whose donor budgets are subject to unpredictable change and whose decision-making is much less influenced by personal factors. Wambugu's decision to maximize profits while seeking to begin serious work on the retail market (even at smaller margins) seems, for the moment, the only practical expedient, since the very large margins available to successful producers will inevitably attract competition, most likely in the medium term from well-established formal

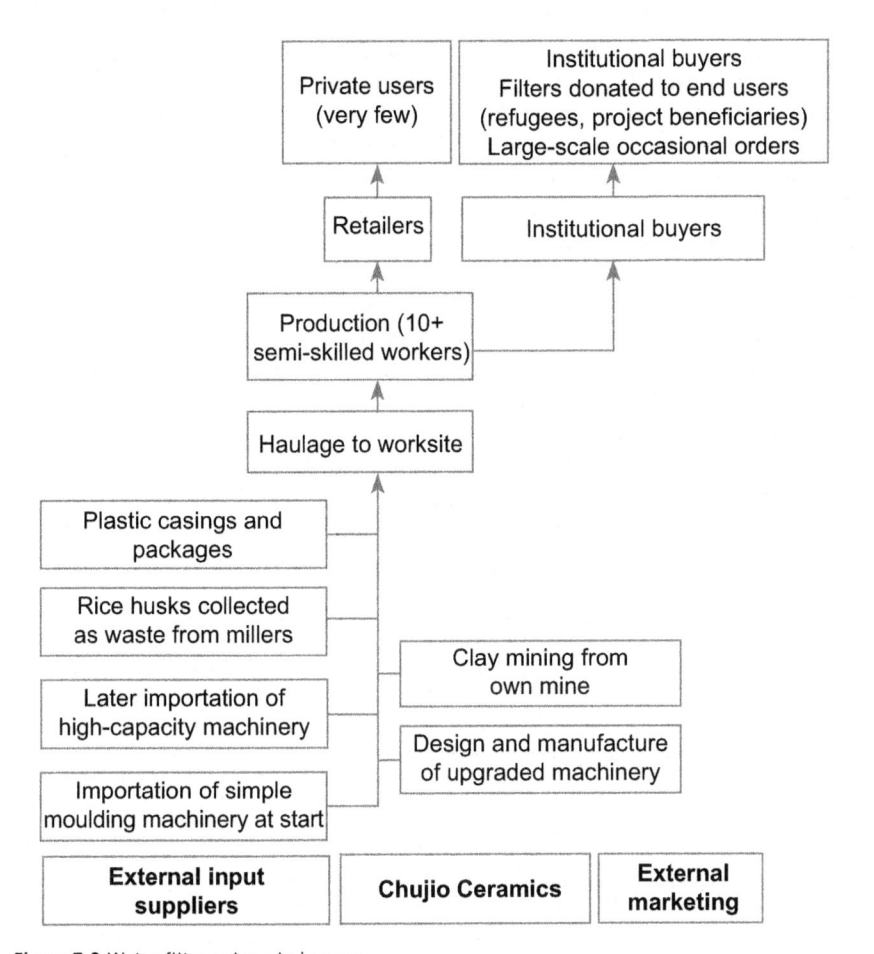

Figure 5.3 Water filter value chain map

sector players. Diversifying further into high value-added areas is therefore a parallel strategy that the company is pursuing as it continues to upgrade its technological base.

Impact

Chujio's effect on the market and contribution to energy savings in Kenya should not be underestimated, with household and environmental savings massively outweighing its initial $3,000 (1986 dollars) investment cost. It is important to recognize that Chujio is not the sole producer, but supplies about 50–60 per cent of the market. Without Chujio, this market would certainly have emerged, but at a much slower pace over an extended period of time. By raising the bar on product quality and customer service it secured the KCJ's future and more rapid acceptance in the market. The Jiko has had a much tougher time in other countries where it has been introduced, mainly because the materials used were often not tested and the production technology was basic. These problems were eliminated in Kenya by Chujio. What, then, has been the economic and environmental impact?

The immediate beneficiaries of Chujio's liner work are about 100–150 liner-makers who buy from the satellite network, and whose income from the KCJ is estimated to be roughly 50–70 per cent greater than when making the traditional charcoal stove. More important, these margins create the incentives that allow a market of about 350,000 consumers to be serviced, each of whom saves between $150–200 per year in energy costs.

Thus, the real story and multiplier effect is amongst the users of the stove, whose savings in aggregate are several thousand times greater than the original investment, not to mention the impact on the environment. As Salehuddin Ahmed of BRAC has said, 'Small is beautiful, but big is necessary'.

Tables 5.2 and 5.3 suggest the aggregate impact of this project on the environment and on household incomes, annually.

Table 5.2 Impact of Chuijo Jiko on household incomes

Item	Amount
Estimated number of standard Jiko liners produced by Chujio annually	175,000
Average life of a Jiko liner (years)	2
Total Chujio Jiko-lined stoves in use (assuming two-year liner life)	350,000
Estimated financial savings per family per year	$173
Estimated total annual financial household savings	$60,550,000

Table 5.3 Impact of Chuijo Jiko on the environment

Item	Amount
Average charcoal saved per stove per year	240 kg
Total charcoal saved per stove per year by Chuijo-lined stoves	84,000 tonnes
Weight of wood to produce 1 kg of charcoal	7 kg
Total reduction in biomass demand owing to Chuijo stoves	588,000 tonnes
Sustainable annual biomass harvest: 1 hectare managed forest	4 tonnes
Nominal annual land area 'saved' by Chuijo liners	147,000 hectares

Conclusions

There are many lessons to be drawn from Chujio's experience. The product has to deliver the right level of performance and quality, satisfy large-scale demand and be reasonably priced. This does not mean it needs to be cheap. Planners and promoters of technology need to recognize that production needs to be as efficient as possible, focused on the prime result, which is to satisfy a mass market with a product that delivers what it claims and does not break down. Technology is the key to stabilizing a strong presence in the informal sector – and escaping from it. It needs to be 'absorbed' by the manufacturer in a way that turnkey technology transfer cannot achieve and indeed, may undermine. This requires capabilities and aptitudes that are unusual and fit no simple profile. Early dependence on credit is one of the least effective ways of encouraging microenterprise growth in the informal sector, because it increases risk at just the time that a young enterprise is engaged in its most intense learning and guesswork. Without the protection and recognition afforded the informal sector by the government of Kenya, none of this would have happened.

Wambugu is an unusual man. His capacity to innovate, respect his customers and master new technology, allied to his willingness to defer consumption in favour of reinvestment, has made him a game-changer in Kenya's Jiko industry. His natural understanding of how to build and manage a value chain, while acting in ways that are consistent and fair, has built him the trust that has enabled his core business to thrive. But his ability to master changes in the technological base of his business has been key to building a strong competitive position and allowed him to evolve into the consummate technician and entrepreneur that he is today.

Reference

Global Alliance for Clean Cookstoves (2013) *Kenya Country Action Plan 2013*, Washington, DC: Global Alliance for Clean Cookstoves.

About the author

Hugh Allen has worked in development since 1970 and is best known for his work on savings groups. This article harks back to an earlier time in his work when he was involved in technology transfer in East Africa. It was through this experience and that of working with ATI that he became part of the Jiko stove story in Kenya.

CHAPTER 6

Granite in Odisha – from Indian quarries to European kitchens, if government allows

Malcolm Harper

Abstract

This case describes the value chain for polished granite, from quarries in eastern India to users worldwide. The data were obtained through personal on-site interviews and from local informal informants. The case demonstrates that a potentially dangerous and environmentally damaging value chain can employ and in some cases substantially benefit disadvantaged people who have few other opportunities, and it also shows that clumsy and ill-managed state intervention can seriously injure such value chains and those who work in them.

Keywords: granite; quarrying; Odisha; Orissa; India; stone polishing

The industry – worldwide and in India and Odisha

Granite is a very popular stone for high-quality exterior building surfaces, for memorials and for flooring, kitchen and bathroom surfaces. In earlier times granite was quarried, cut and polished near to where it was used. Aberdeen in Scotland, which is still known as the 'granite city' because many of its buildings were built with the stone, was home to large numbers of granite quarries and polishing factories including Rubislaw quarry, opened in the early eighteenth century and closed in 1971 and believed to be the largest man-made hole in Europe.

Granite remains popular, and the annual international trade in granite is increasing at an annual rate of between 10 per cent and 15 per cent. In spite of its relatively low value per tonne, which makes it expensive to transport, most granite that is used in the so-called 'developed countries' of the world is imported from poorer countries. The United States is the world's major user of granite, and there are many places where granite can and has been economically quarried in North America. In 2011, however, 87 per cent of the USA's granite requirements were imported. Over a third came from Brazil, about a quarter from China and 18 per cent from India (basicsmines.com, 2010).

The quarrying and processing of granite involves hard and dangerous work, it can destroy whole landscapes and ruin agricultural land, and it uses

http://dx.doi.org/10.3362/9781780448671.006

a great deal of water and produces large quantities of scrap stone, dust and other pollution. The transport of the raw stone blocks to processing units and ports also causes severe damage to roads. Hence, like many similar products, its production has in large part been 'outsourced' to poorer countries where labour and other standards are less demanding, and where such regulations as do exist can often be circumvented.

India exports around 5 million tonnes of granite per year. The states of Karnataka, Gujarat, Jharkhand and Rajasthan are responsible for almost 1 million tonnes each; Andhra Pradesh, Madhya Pradesh and Odisha (formerly Orissa) export about 200,000 tonnes each; the remaining states account for relatively insignificant amounts. This amounts to about 13 per cent of world trade, in tonnes; the data are for raw and polished stone together, but Indian exports are mainly rough unprocessed blocks, which are exported to Taiwan, China and elsewhere for processing and further re-export or local use. Indian production of granite for local consumption is estimated to be around the same volume, but tends to be of lower priced and lower quality material.

According to *Stone and Tile in India* (2014), Odisha is the world's main centre for the production and export of marble stone. There are 20 different colours of granite in Odisha, and four of the world's five most attractive colours are available there; lavender blue in Berhampur district, ikon brown in Paralakhemundi, sweet green in Titilagarh and black in Koraput. In Paralakhemundi in particular there are said to be abundant reserves of high-quality granite.

Problems in India and in Odisha

As is often the case in India, and in particular in industries that are affected by land access rights and environmental issues, the industry is strongly critical of government, but also expects government to take steps that would elsewhere be undertaken by the industry itself. These issues often arise as a 'hangover' from the earlier so-called 'licence raj', where just about every important business decision, such as expansion, or closure, or export, had to be submitted and approved by government. This situation still prevails in some areas, particularly in less developed states such as Odisha.

Many granite polishing units are in difficulty because they cannot access sufficient supplies of raw stone; this is partly because many granite quarries have been closed down by government, for alleged infringement of environmental or other regulations, and partly because competitive polishing units in the nearby state of Andhra Pradesh, which are often closer to the quarries than units in Odisha itself, are able to offer better prices and more regular orders.

The Odisha industry's response to this is generally not to offer better terms to quarry owners, or to help them improve their operations, but to ask the state Department of Steel and Mines to intervene, and to ensure that Odisha polishing units have priority access to supplies from quarries in Odisha. They accuse the quarry owners, many of whom come from outside

Odisha, of selling only low-grade material and off-cuts within the state, and of diverting the higher-quality material to granite polishing units in Andhra Pradesh, Karnataka and Tamil Nadu. They have demanded that the state government should open a 'granite park' for polishing units, with subsidized power and other facilities, and they have also asked the government to double the royalty which is charged on sales of raw granite blocks to out-of-state purchasers.

The Hindu newspaper reported on 26 June 2012 (*The Hindu*, 2012) that forest land in Ganjam district of Odisha, adjacent to Paralekhamundi, had been illegally leased out for granite quarrying, and that transport permits for granite blocks and operating licences had been illegally provided. The operations of some quarries were also said to be polluting a major canal that supplies drinking water to Berhampur, the central town of the district. In 2013 the same newspaper reported a claim by the local opposition member of parliament that illegal dumping was causing landslides, which had killed four workers, and that protected forest trees were being cut down to make way for quarry operations. The state was said to have lost almost US$120 million in unpaid quarrying royalties. *India Today* (2012) also reported that 215 quarry operators in Odisha were taking advantage of a regulation that allowed applicants for licence renewals whose applications were not dealt with within three months to be automatically granted an extension. All this was part of the maladministration of the state, which has paradoxically lead not only to misuse but also to underuse of larger-scale mineral resources, such as iron ore, bauxite and coal, as the whole system of regulation of mineral exploitation is such as to discourage legitimate national and international operators.

The granite industry also experiences regulatory problems in other states. In Andhra Pradesh 26 quarry workers were killed in 2010 when badly stacked material fell on them (Newswala.net, 2013) and in Tamil Nadu in December 2013, 15 quarries and polishing units near Bangalore were ordered to close within three months because of their alleged illegal operations along a sensitive riverside.

Granite quarrying is difficult, dangerous and dusty, and cutting and polishing are also hazardous. The large slabs and blocks, which can weigh up to 50 tonnes, cause many injuries. A single slab often weighs 500 kg or more; these slabs are held by crane, operated with clamps that grab the centre of the slab. If the slab breaks, it can easily crush workers' limbs or swing out of control and kill them; very few workers in Indian granite quarries or polishing units appear to wear protective helmets or boots, and many of them work in sandals or even bare feet.

There have been some attempts at the importing end of the value chain to improve working conditions, and in particular to avoid bonded labour and the employment of children, who are often used to carry scrap material away from working faces in quarries, and to manipulate slabs on polishing machines. Pressure on this issue has been brought to bear on granite processors

62 COMMERCIAL AND INCLUSIVE VALUE CHAINS

in Aberdeen such that, since most of its own quarries were closed, it has become a major centre for importing granite, in particular from India. In 2007 a spokesperson for Anti-slavery International stated:

> India uses bonded labour to quarry granite and slate and British firms are contributing to that. They must ensure their supply chain is free of slave labour. Bonded labour is the most extensive form of slavery and is prohibited under international and Indian law. But it's widespread (twocircles.net, 2013).

The Aberdeen-based granite processing company Kirk Natural Stone, which imports granite from India, responded:

> We try to ensure that everything is ethically sourced but what can you do from this distance? You can go and see your supplier but you have to take what they show you at face value. I don't believe pregnant women and nine-year-old boys are working at the quarries we use (twocircles.net, 2013).

The Indian Committee of the Netherlands (2006) claims that 90 per cent of Indian quarry workers are bonded to their employers by advance payments of some kind; this can of course mean anything between a few days advance of salary and a commitment that can last many generations. As is so common in India, this is forbidden in law, but ill defined and very common in practice. One of our respondents told us that she had received a substantial advance of several months' wages when she started work in a quarry in 1999. She used the money to build a home for herself, and did not regard it as unfair or exploitative. No working children were observed in our own visits; this does not mean that none were employed anywhere in the industry.

In spite of all these problems granite quarrying and processing continues in Odisha, albeit on a smaller, less formal and probably less efficient scale than might be possible were the industry properly regulated. The State Ministry of Micro, Small and Medium Enterprises itself reported in December 2013 (IPICOL, 2013) that a total of 329 quarrying leases had been granted in the state, 118 in Gajapati and Ganjam districts, where Paralekhamundi is situated.

This case study attempts to assess the condition of the granite value chain in Odisha by examining the operations of two quarries and two polishing units in the neighbourhood of Paralekhamundi. This is not a comprehensive or representative study, but it is hoped that the findings will provide some useful objective, albeit anecdotal, data.

The granite value chain

Figure 6.1 summarizes the main features of the value chain for what is known as 'dimension' granite, that is, stone that is cut to size rather than supplied as gravel or 'chips' for roads and similar construction purposes. The rupee prices are approximate estimates of the rupee price per cubic metre at each stage. The approximate exchange rate at the time of writing was Rs60 per US$1. These

GRANITE IN ODISHA – QUARRIES TO KITCHENS   63

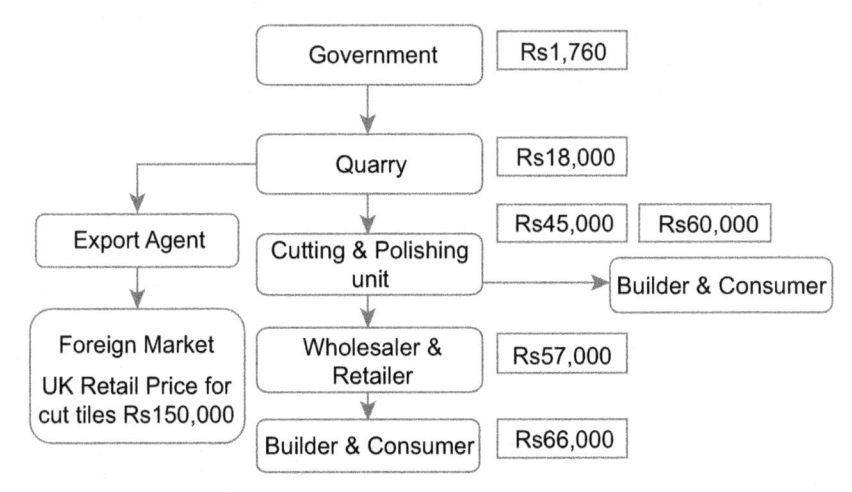

Figure 6.1 Main features of the granite value chain, per tonne in Indian rupees

prices and other information were obtained from observation and informal conversations 'on-site' with the owner and workers at Meenakshi Export, a quarry near Buguda village, which is about 10 km from Paralekhamundi, and with Baba Granites, a granite polishing unit in the town; the figures vary widely depending on the variety and quality of stone, market conditions and the type and extent of processing.

The figure for the government is the royalty per tonne paid to government by the quarry operator for each tonne of granite extracted. The remaining figures, all of which are in Indian rupees, show the price charged to each player by the preceding supplier in the chain, which covers the cost and any profit. The United Kingdom retail price is calculated on the average price charged for 10 millimetre thick floor tiles; the price for 30 mm thick kitchen surfaces is around double this level, because higher-quality stone is used and floor tiles are produced in standard sizes whereas kitchen counter tops are cut to order.

The UK distributor of kitchen surfaces obtains 30 mm thick slabs of about 3 m by 1.8 m, ready polished, cuts them to size and bevels and polishes the edges as required. These are bought from specialist importers, and the distributor is not always aware of their source; one importer supplies what is known in the trade as 'Orissa blue' granite, but he was not himself even aware that Orissa (now Odisha) is in India.

The quarry

The government's role is to decide where quarrying is to be permitted, and then to issue licences. Most quarries are in hilly and forested areas, and most such land, and all mineral rights in India, belongs to the government. Licensing is inevitably difficult, because of the different interest groups who are involved; local people want employment, but farmers do not want to lose what is effectively free grazing land, and 'minor forest produce' such as firewood, medicinal herbs, leaves for making leaf plates and other items are an important source of livelihood for the so-called 'tribal' people who are descended from the original pre-Aryan inhabitants of the Indian peninsula and who generally remain a marginalized sub-caste group outside mainstream society.

Once a site has been selected, an open bidding process is undertaken, at least in theory, whereby those who wish to quarry the site bid for what is usually a ten-year licence. They may also have to construct or strengthen the access roads to the site, and to undertake, again in theory, to restore the site to its previous condition at the expiry of the licence. The capital payment is often quite nominal, and may be based on very out-dated values, such as the Rs2,500 (around US$42) application fee that is required. The annual fee is charged per tonne of granite removed from the site; the Meenakshi quarry initially paid Rs1,760 per cubic metre, as shown above, but this is increased by 40 per cent every three years, that is twice during the life of the lease.

Like most modern quarries in India and elsewhere, the Meenakshi quarry does not use blasting to extract stone. Large 45-tonne blocks, of around 2 m by 2 m by 3 m, are cut from the exposed rock faces using diamond-tipped drills and cutting ropes that are electronically controlled, and lubricated by a continuous flow of water or kerosene to prevent overheating. The blocks are tipped forward from the face by bulldozers and manipulated on to trucks to be carried to the processing units.

Mr Y.V. Rao, the proprietor of the Meenakshi quarry, hires a specialized team of skilled workers from Rajasthan to drill and saw granite blocks. They each receive about Rs1,000 per day, as does the bulldozer driver, and the company provides the bulk of the equipment. The total cost of the cutting machinery alone is around Rs1 million, and the bulldozer costs as much again. The running costs include 200 litres of diesel every day, kerosene for lubricating the wire rope saw, hydrated lime imported from China, which is used during drilling, as well as labour and the government royalty. The quarry's total cost is on average around Rs18,000 per cubic metre. The blocks are sold in their rough form for around Rs25,000 per cubic metre, depending on colour and quality, to customers in Taiwan, China and Germany, and to processing units in Srikakulam in neighbouring Andhra Pradesh, one of which is also owned by Mr Rao. He is under pressure from the Odisha state authorities to add more value locally, and plans in due course to open his own polishing unit adjacent to the quarry.

Some subsidies are available for local businesses such as Meenakshi, but Mr Rao feels that the business is reasonably profitable in its own right and it may in any case not be worth the 'hassle' involved in obtaining this assistance, particularly as he is himself from Andhra Pradesh, not from Odisha.

The Meenakshi quarry employs 20 experienced male workers who are paid Rs240 per day, and four local women paid Rs160 a day to bring drinking water to the workers. This compares to the Rs126 a day that is paid under the

Indian government's National Rural Employment Act (NREGA). Every rural household in India is entitled to 100 days work a year under this Act, and NREGA wage rates are often considered a 'floor' rate for unskilled workers. The quarry operates six days a week for eight hours a day.

Quarry workers' remuneration and conditions of work

Ten workers at the Meenakshi quarry answered the PPI Progress out of Poverty questionnaire (http://www.progressoutofpoverty.org/country/india), for the day when we met them, in early December 2013, and for 2008, five years earlier. Their average score in 2008 was 18, and 6 of them scored under 20, which indicates that the likelihood of their falling below the income poverty line of $1.25 a day was over 60 per cent. The average for their 2013 responses was 31, and none scored under 20; this does not mean that they were wealthy, but the likelihood of their falling below the $1.25 poverty line was reduced to under 30 per cent.

This is significant, and the improvement in these workers' conditions appears to be above the 50 per cent increase in state product per head that is said to have been achieved in Odisha over the same five-year period, and rather further ahead of the 37 per cent improvement for Gajapati district (Government of Odisha, 2013). These are not statistically valid comparisons, but they do confirm the view of many of the workers that they are better off than before.

P. Raju is 34 years old. He operates a drill at the Buguda Granite quarry. He started work there in 2009; he earns Rs240 per day to support his wife and their two children. He is satisfied with the way his household's condition has improved as a result of his job. His employer provides him with lunch and drinking water, and what is most important is that the work is continuous; he is employed full time, every day, every month, and the quarry is within easy reach of his family's home in Buguda village. Before he got this job, he used to have to migrate here and there for occasional work; now he has a steady job, close to home.

T. Murali Reddy is also a labourer at the quarry. Since he started work there he has been able to afford a TV, a video player, a mobile phone, a steel cupboard and a bicycle. He often earns a bonus on top of his regular wage of Rs240 a day, and although it is hard and dangerous work, he now feels that he can support his wife and his two young sons.

Two other labourers were less satisfied. Vijayalakshmi has been breaking rocks into small stones in granite quarries since 1999. She started work at the age of 17, and received an advance of Rs10,000, which she used to build a home for herself. Her wage has gone up from Rs100 to Rs200 during this period. She has two sons; the younger one is at school and the 15-year old works in a textile mill. It takes her five days to fill a truck with broken stones, and she dislikes the job because it hurts her hands, but she carries on because she does not know about any other work.

Appalaraju is 29 years old. He dropped out of school aged 11 and has been breaking stones ever since. He has worked in ten different granite quarries in Tamil Nadu, Kerala, Karnataka, Andhra Pradesh and Odisha. He told us that the work was very, very hard when he was so young, but there was no other way. His parents were already working in the quarry; his three brothers and two sisters have grown up together in the quarries. He breaks rocks into medium sized blocks that are used to make walls for houses, and earns Rs3.50 for each piece. He produces between 100 and 150 pieces each week, and feels as if his life has come to an end. His only hope is to see his children through to higher studies so that they can have a better life than he has.

Cutting and polishing

The rough blocks of granite are transported from the quarries to the processing units that belong to the quarry operators, or are independent, and are generally located outside the hilly areas where they are quarried, on main roads and nearer to urban areas. The blocks are cut into slabs of the required sizes on large computer-operated horizontal sawing machines, then polished in several stages using specialized grinding machines and abrasive metal or silicon carbide 'bricks', along with large quantities of water or kerosene to lubricate the cutting surfaces and to remove the waste. It requires highly skilled operators; it is important to ensure that the surfaces are not damaged, particularly when they are at the final stages of processing.

Baba Granites is one of the larger polishing units in Paralakhamundi. The unit was started by Sujata Sahu and her family in 2002 in order to process granite from the family's existing granite quarries, but their quarries have recently been closed by government order, along with many others. There used to be about 20 quarries in the neighbourhood of Paralakhamundi, employing over 500 people; the average monthly income for skilled workers was around Rs10,000 plus bonus, or Rs7,000 for unskilled labour, but 18 quarries were closed in 2010 and only two remain, including Meenakshi. The government's annual royalty from these quarries used to be about Rs20 million; this has been reduced to Rs4 million.

Some of the quarry workers who lost their jobs were migrants from Andhra Pradesh and have returned home. Others were from local villages close to the quarries, like those from Barguda who still work at the Meenakshi quarry. Those whose jobs have disappeared have suffered a major drop in their living standards. Some have migrated to nearby cities for casual labour work, and others are eking out a living by peddling a rickshaw, or as vendors or by working on other people's farms; some have accepted work under the NREGA scheme, which is intended to provide a safety net for the destitute. The quarry closure has been a major blow to the local economy.

One cubic metre of rough granite can be cut into as much as 60 square metres of granite slab, depending on the thickness. Cutting and polishing costs about Rs500 per square metre, so that the total cost of cutting and polishing 1 cubic metre comes to around Rs18,000. The transport cost per cubic metre is Rs3,000, and the wholesale selling price per square metre of polished slab comes to around Rs1,200. The final retail selling price in India is around Rs1,600 per square metre.

Odisha has many colours of granite, and prices per square metre vary widely depending on the colour and the slab size. The most popular is (or was) ikon brown; this fits well into many modern design schemes, but the quarries that have been closed were the major sources and Baba Granites and others have to use a special dye on grey granite in order to achieve a similar colour. This is costly, and although it is soaked deep into the material and is said to last at least 20 years, it is clearly not the same as 'the real thing'. The supplies of genuine Odisha stone are limited but, as we have seen, a large quantity of what is quarried is sold to polishing units outside the state. It is sold as 'Orissa blue' and so on, but there is of course no guarantee as to the origin of the original material. Odisha's position in granite in relation to India is not dissimilar to that of India and the rest of the world; the major value addition process takes place elsewhere.

Granite processors' remuneration and work conditions

The skilled cutting and polishing workers earn around Rs400 a day; unskilled labourers get about Rs300, which is substantially higher than the quarry workers because the units are closer to other jobs. The labourers who move the stones and slabs from one machine to another, and must thus manipulate the heavy lifting tackle and cranes, earn around Rs300 a day.

A total of 14 workers at Baba Granite processors and at Odisha Granites, a neighbouring processing unit, answered the PPI questionnaire, for early December 2013, and for 2008, five years earlier. Their average score in 2008 was 23, well above the figure of 18 for the quarry workers, and the average for their 2013 responses was 38, as opposed to 31 for the quarry workers.

This suggests that these workers were significantly better off than the quarry workers at both periods; the work in both places is physically hard and hazardous, and only a few of the workers appeared to use protective footwear, goggles or helmets. Workers dislike such equipment, particularly in hot places, even when it is supplied free of charge, because it is uncomfortable and can actually make it hard to respond quickly to dangerous situations.

Nevertheless, the workers at the processing units seemed in general to be happier than those at the quarry. Some of them work on a piecework basis, which enables them to increase their earnings as their skills improve. At Laxmi Granites, the polishers earn Rs3 per polished sheet and can polish between 70 and 100 sheets a day. This enables them to earn between Rs7,000 and Rs9,000 a month, which is well over the earnings at the quarries.

Rakesh is 26 years old. He used to work in a quarry but when it closed in 2010 he shifted to Odisha Granites. He is happy with the change. He used to earn around Rs100 a day at the quarry and now earns Rs250 a day as a polisher and helper. His job is under cover, rather than being exposed to all weathers at the quarry, it is less hazardous and he is learning new skills. He plans to learn as much as he can about cutting stone and hopes in two years to move to another company as a full-time polishing machine operator. He cycles to work from a nearby village, and can now help to support his two younger brothers, one of whom is studying for a degree at college. Rakesh has also been able to help his sister to get married.

Wholesalers and further members of the value chain

Annapurna Marbles purchases bulk quantities of cut and polished granite slabs from processing units such as Baba and Odisha Granites and holds stocks so that builders and others can obtain what they need at short notice. The prices differ according to the colour, size and thickness required. These figures are quoted in rupees per cubic metre, but they are actually for granite slabs, cut to size and polished.

A typical standard granite slab costs around Rs60,000 per cubic metre in India when cut and polished into slabs which are ready for installation. This material is sold to the final customers, or more often to the contractors who in turn supply it and fit it for the actual users, for around Rs66,000 per cubic metre of granite.

As was shown in Figure 6.1, the same or similar product sells in the United Kingdom for the equivalent of Rs150,000 per cubic metre when it is ready for installation. The difference between this and the Indian price of Rs60,000 is in part accounted for by transport and other shipping costs: to the port in India and from the UK port, perhaps trans-shipped from a continental European port to the importer's warehouse, sometimes thence to a specialist firm that carries out the final cutting, edging and finishing operations to the customer's specifications, and finally to the place of installation.

References

basicsmines.com (2010) 'Outlook Reexamined for 2009: "Early Bird"', Issue W. Online at <www.basicsmines.com> accessed 28 December 2013.

Government of Odisha (2013) *Economic Survey of Odisha, 2011–12*, Bhubaneswar: Government of Odisha.

The Hindu (2012) 'Granite Quarry Opposed', *The Hindu*, 26 June.

Indian Committee of the Netherlands (1996) *Quarry to Graveyard*, Utrecht: Indian Committee of the Netherlands.

India Today (2012) 'Granite leases – mining money', *India Today*, 19 December.

IPICOL (Industrial Promotion and Investment Corporation of Odisha Limited) (2013) 'Granite', online at <http://ipicolorissa.com/index.php?option=-com_content&view=article&id=29:granites&catid=7:minerals&Itemid=24> accessed 27 December 2013

Newswala.com (2013) 'Prakasam Dist.: 20 feared killed in Andhra quarry mishap', online at <www.newswala.com> accessed 29 December 2013

Stone and Tile In India (2014) 'All India Granites & Stone Association: Promoting Indian Stone Industry', 6 January, online at <http://stonetileindia.com/?p=1569> accessed 20 January 2014 twocircles.net (2013) 'Scotland's "granite city" imports stone from India', online at <http://twocircles.net/2007may03/scotlands-granite-city-imports-stone-india.html#.VISGm9KUfkc> accessed 29 December 2013.

Acknowledgements

Material for this case study was gathered with the assistance of the following MBA students from Centurion University, Paralekhamundi campus: Ratnesh Chandra Behera, Santunu Biswal, Debashis Dakua, I.B.K.Vardhaman Gupta, Sabir Kumar Anand, Sunil Kumar Sahu, Abinash Padhy and Sunil Kumar Dakua. Their faculty supervisor was Dr Umakanta Nayak. Their assistance is gratefully acknowledged.

About the author

Malcolm Harper is Emeritus Professor of Cranfield University in England, and has taught at the University of Nairobi and Cranfield University; he has worked since 1970 in enterprise development, microfinance and other approaches to poverty alleviation, mainly in India.

CHAPTER 7
Remittances – from the global diaspora to the poor in Somalia

Abdi Abokor Yusuf

Abstract

This case describes the system whereby money is transferred from the Somali diaspora in Europe, the United States and elsewhere to needy people in Somalia and Somaliland. It demonstrates how a low-cost multinational business of Somali origin has developed a value chain for remittances from the international Somali diaspora to family and friends remaining in Somaliland. The chain brings badly needed assistance to needy people and involves large numbers of small independent businesses in the remitting as well as the recipient areas. The value chain is under threat from overzealous and irrelevant anti-money laundering restrictions by a large British-based bank.

Keywords: remittances; Somalia; Somaliland; Dahabshiil; Barclays Bank; cash transfers

Remittance companies send money from around the world to Somalia, and are a crucial source of income for recipients. Dahabshiil is based in Somaliland and is Africa's largest money transfer business. The transfers are a vital source of income for many of Somalia's poorest people.

Somali remittance companies serve the worldwide Somali diaspora by enabling them and others who live outside Somalia to transfer money to those who remain in Somalia. As well as being used for daily consumption, and thus reducing poverty, the remittances help the reconstruction and rehabilitation of the country, allowing exporters, wholesalers and traders to access finance for their businesses. The main function of remittance companies is money transfers, but because there is at present no operating banking system in Somalia they also provide other functions such as current and deposit accounts and credit. They also service individuals, private enterprises, government and international development organizations involved in humanitarian and development activities in the country. These agencies assist in development and also create jobs. In Somaliland they are said to employ 4,000 people and annual remittances by the diaspora are estimated to exceed US$500 million,

http://dx.doi.org/10.3362/9781780448671.007

according to Dahabshiil (Dahabshiil.com, 2014), Somalia's leading money transfer company (Government of Somaliland, 2012: 97).

Remittances are a major source of livelihood for their recipients. Studies in Hargeisa and Burao in Somaliland and in Bosasso in Puntland suggest that remittances make up nearly 40 per cent of urban households' income and roughly 14 per cent of rural incomes (Government of Somaliland, 2012: 97). They are a lifeline without which dependence on international food aid or large-scale starvation would be unavoidable. In 2005 Somaliland received $105 per capita in remittances, which contributed a fifth of Somaliland's gross domestic product. This ranks fourth among remittance-dependent economies after Togo (38.6 per cent), Lesotho (28.7 per cent) and Jordan (23 per cent) (Shire, 2004).

Dahabshiil is Africa's largest money transfer business. They have 24,000 agents and branches in 144 countries and have been operating for over 40 years (Dahabshiil.com, 2014). It is not difficult to transfer money with Dahabshiil. Customers who wish to send money to Somalia register with a nearby agent, they hand over the money they wish to send and it is confirmed to be ready for collection by the recipient within 15 minutes. Customers can check the status of their transaction online and the recipients are notified either by phone or text that the money is ready for collection.

There has been mass migration from Somaliland since 1980 as the country has suffered from protracted civil war. People have migrated to find safety and economic opportunities in Europe, America, Australia and the Gulf. They support their families back home by sending a significant portion of their income to their dependents; this significantly reduces poverty and food insecurity in the country.

Remittances from migrants help education by investing in schools, paying teachers' salaries and providing facilities such as books and computers. Migrants compete with one another to develop their communities back home. 'Children in the households of people who receive remittances have better school attendance rates, and the remittances encourage families to educate their children. Sibling solidarity plays a particularly crucial cultural role in the education and welfare of children and young people' (Maimbo, 2006: 2).

This study was conducted in Hargeisa, the capital of Somaliland. Figure 7.1 illustrates the value chain for the remittances, and is followed by Table 7.1 that shows how the 5 per cent transfer fee is allocated between the various links in the chain.

Dahabshiil has accounts with a number of commercial banks worldwide, and Barclays was the most important intermediary in the UK and Europe. In 2013 Barclays notified Dahabshiil that this vital service was to be withdrawn, presumably because the bank was afraid that it might be accused of money laundering and face a multi-billion dollar fine such as was levied on some of its competitors. Dahabshiil took the case to law, and won an injunction that stopped Barclays from immediately withdrawing the service; since that time Dahabshiil has made arrangements with other service providers.

REMITTANCES TO THE POOR IN SOMALIA

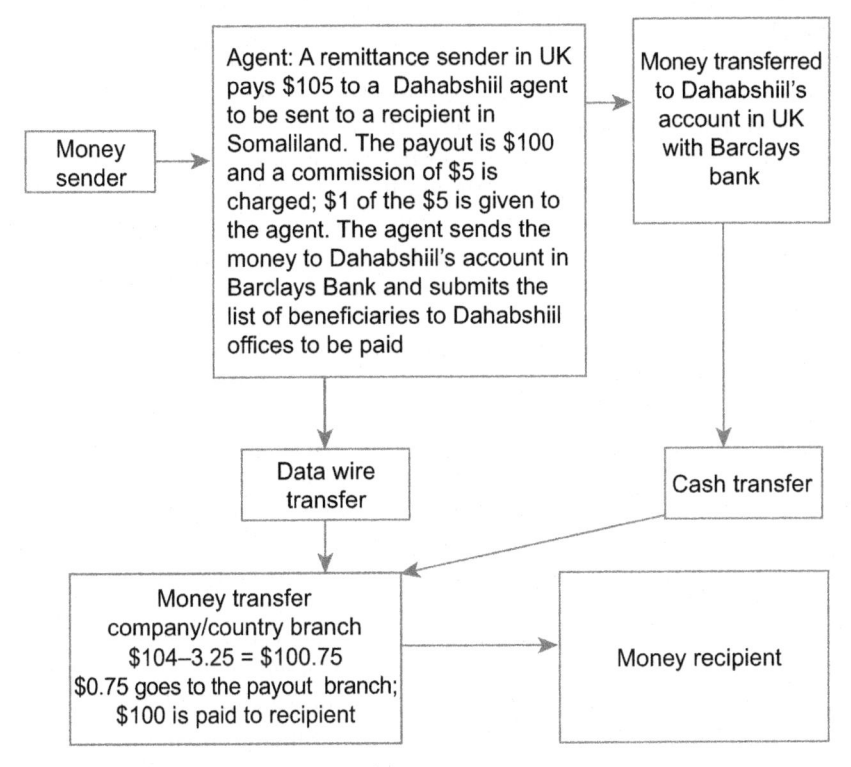

```
                    ┌─────────────────────────────┐
                    │ Agent: A remittance sender  │      ┌──────────────────┐
                    │ in UK pays $105 to a        │      │ Money transferred│
                    │ Dahabshiil agent to be sent │─────▶│ to Dahabshiil's  │
    ┌──────────┐    │ to a recipient in Somaliland│      │ account in UK    │
    │ Money    │───▶│ The payout is $100 and a    │      │ with Barclays    │
    │ sender   │    │ commission of $5 is charged;│      │ bank             │
    └──────────┘    │ $1 of the $5 is given to    │      └──────────────────┘
                    │ the agent. The agent sends  │
                    │ the money to Dahabshiil's   │
                    │ account in Barclays Bank    │
                    │ and submits the list of     │
                    │ beneficiaries to Dahabshiil │
                    │ offices to be paid          │
                    └─────────────────────────────┘
```

Figure 7.1 Overview of a Dahabshiil money transfer process

Table 7.1 Value added along the chain for $100 remittance to Somaliland

Stage	Amount (US$)
Sender in USA, UK, etc. pays to one of Dahabshiil's agents	105
Agent deducts $1 commission, remits $104 balance to Dahabshiil's bank in UK	104
Dahabshiil sends list of payees to branches in Somalia, and its UK bank remits total receipts daily to Dahabshiil's bank account in Somaliland, via Dubai	–
Dahabshiil deducts fee of US$3.25, and remits balance to the respective branch	100.75
Branch deducts commission of 75 cents, and remits balance to recipient	$100

The money transfer companies have outlets throughout Somalia. The process of sending or paying money is carefully documented. The details of both the sender and the recipient are recorded for each transaction, including the name of the client, a copy of their identification or passport, their address or their occupation. If the client is not recognized by the teller, he or she can be asked to name his clan. This is a good tracing mechanism for Somalis, and in case of doubt a person can be asked to give a reference of an acquaintance or friend. This form of traditional recognition is still used, but the agents also rely on passports and other more formal means of identification. In effect, everyone can be identified, even those with no fixed residence.

Until recently the clan system was also the most effective way to inform recipients that a remittance had arrived, but modern information technology is also used. Clients are informed by SMS messages if their mobile numbers are in the system. The clan identification system is still functioning, however, and is used for new clients who may not already be known to the company's staff.

The paying out offices in Somalia are either agents who are paid on commission or money transfer company staff who receive regular salaries. This depends on the volume of business expected in each locality, but most of Dahabshiil's over 200 outlets in Somalia are company-owned. The commission agents are usually paid about 15 per cent of the commission earned by the company. This covers their costs, and if, as is often the case, the agents are also running retail shops or other businesses from the same location, the money transfer agency helps to generate traffic. The transmission charges are lower for local transactions within the country; a 0.5 per cent is charged on a $100 transfer, for instance, and the rate is lower for larger amounts. Both Somalia and Somaliland have their own currencies, but all Dahabshiil's transactions are in US dollars. There are large numbers of very competitive money-changers in every town, so that dollars can be converted to Somali or Somaliland shillings without difficulty.

There are a number of money transfer companies through which Somali people can move money to their families and others. In addition to Dahabshiil, which is the largest and longest established, there are nine other significant companies: Jubba Express, Amal Express, Hodan Global, Kaah, Amaana, Mustaqbal, Tawakal, World Remit and Iftin. There is also a local mobile money transfer company called Zaad, which provides mobile wallets, through which money can be electronically transferred. So far, this service only operates within Somalia.

Most of these companies charge similar rates, but all are less expensive than the major international competitors. As of August 2013, the cost of sending $200 to Somalia, paid out in local currency, was $10 through Dahabshiil and $15 through Western Union. Western Union and the other international companies have good networks of paying-in agents worldwide, but Dahabshiil and other local firms also have excellent agent coverage in all areas in the UK, the USA and wherever there are concentrations of Somali immigrants. However, the Western Union website (http://locations.westernunion.com/search/somalia/woqooyi+galbeed/hargeisa) showed only one location in Somaliland, in Hargeisa, and none in the rest of Somalia. Dahabshiil has over 200 of its own offices and agents throughout Somalia, and its competitors have many more.

Dahabshiil is itself a classic 'rags to riches' story. It is still a family business, and the present chief executive officer is Abdirashid Duale, the son of the founder. The company employs nearly 5,000 people in over 144 countries, it has corporate offices in London and Dubai, in addition to its head office in Hargeisa, and provides money transfer services to the United Nations, Oxfam, the Department for International Development (DFID), Development Alternatives Inc (DAI), Save the Children and many humanitarian institutions. It is arguably one of the most important multinational businesses in Africa, through its provision of vital money transfers to Somalis and others throughout Africa and beyond.

Dahabshiil was founded by Mohamed Saeed Duale, the present chairman, in 1970. He started as a remittance broker, importing goods from the Gulf States to Somalia for the families of migrant workers, at the request of their relatives overseas. Dahabshiil opened its first shop in Burao, the capital of Togdheer province in northwest Somalia, which is now known as Somaliland. The business grew over the next 18 years, and became the leading remittance broker in the Horn of Africa.

In 1988, the business collapsed as civil war broke out across Somalia. This forced half a million Somalis to flee all over the world. Mohamed Saeed Duale was also forced to leave Somalia and to take refuge in a refugee camp in Ethiopia with his family, but he used his experience and his network of business associates to set up a new remittance venture in Ethiopia, enabling Somali migrant workers to continue sending money to their family members in Ethiopian refugee camps.

The presence of so many Somali refugees outside Africa increased Dahabshiil's business, particularly in the UK, and they set up an office in London. As the UK's Somali population grew, so did Dahabshiil. The company grew steadily, opening offices and agencies in the USA, with particular emphasis on the cities of St Paul and Minneapolis in Minnesota, where there is a large Somali community. In 2009, Dahabshiil launched the first debit card in Somaliland, and in 2010 they opened an Islamic bank in Djibouti. The business has zero debt, it remains entirely family owned and is committed to its fair commission fee policy. Dahabshiil continues to support the Somali community both in Africa and abroad, and has an active corporate social responsibility programme. Five per cent of its profits are invested in community regeneration projects such as schools, hospitals, agriculture and sanitation facilities (Dahabshiil.com, 2014).

The collapse of the Somali government in 1991 destroyed the country's formal banks and opened an opportunity for remittance companies such as Dahabshiil, some of which have developed into global companies. They can compete with banks and other international financial institutions in the global market.

The Somali remittance companies have come up in a harsh environment, but they have enjoyed several advantages. They were able from the outset to secure and operate in areas that were not served by banks, and they have established a comprehensive network of agents and branches to deliver money transfer services. They maintain close proximity to their customers through their physical locations but also through traditional clan linkages.

The transaction costs of the local remittance companies are in general lower than those of their international competitors, and they can provide cheaper, faster and more reliable services, in part because of their clan linkages. Their prices are usually lower than international competitors such as Western Union, but they can also compete through their network and customer care at both the remitting and the recipient ends of the transaction; this cannot easily be matched by their competitors.

Because there are no banks in Somalia, Dahabshiil has also been able to provide other services to their customers, such as saving deposits and investments. It serves ordinary people, including very poor women and others, but also carries out transactions for the government and for development agencies. The Dahabshiil brand is ubiquitous throughout Somalia, and is perhaps as familiar or even more so than internationally recognized brands such as Coca-Cola. Most development aid to Somalia and Somaliland is channelled through Dahabshiil, through which money can be distributed all over the country, to the staff of development institutions but also direct to local beneficiaries when the agency wishes to make cash transfers. The company is large enough to be able to deal with these agencies in their countries of origin, but is also known and trusted by the most humble rural pastoralists in Somalia; this is an enormous competitive advantage. Thousands of Somalis and others use Dahabshiil and its competitors' services every day for personal and business transactions.

Aden Ibrahim of Aalborg in Denmark is a typical personal customer. He has been sending money to his relatives in Somaliland since 1994. He sends money to his father, brothers and sisters and other close relatives who live in Somaliland. The family uses the money to buy food, to pay for utilities such as electricity and water, and for school fees. They consider that the 5 per cent commission is very reasonable. Dahabshiil has several agents in Denmark and Mr Aden uses the agent in Aalborg, which is run by Mohamed Ugas, whom he knows personally. The commission is reduced to 4 per cent or less when he sends larger amounts, such as during the month of Ramadan. His family needs more money at this time since household consumption increases and at the end of the month the Eid Festival is observed and they buy new clothes.

Muna Abdi is a typical remittance recipient. She has four children, and she sometimes receives money from her brother and her two sisters who live in London and in Minnesota. When her money reaches Dahabshiil, she is automatically informed by an SMS message sent to her mobile phone. This usually happens within 10 or 15 minutes of the money being paid to the Dahabshill agent in the UK or USA. She goes to the nearest agent in Hargeisa, where she is recognized by the staff. If she goes to an agent where she is not known, she can show an identity document. After she is recognized by the teller, she signs the receipt and collects the money. She can then buy what her children need for the month and settle any bills she may have incurred.

When Muna Abdi heard that Barclays Bank was going to close the remittance companies' accounts in the UK, she said, 'we will find other options to receive the money, this is not the first time that we hear such information. It started in Minnesota, and it was soon sorted out'.

Mr Mohamoud Adani is a client of Dahabshiil who lives in London; he sends money to his relatives back home in Somaliland. He says that he has two options for sending money home. He can go to Dahabshiil's main office in Whitechapel, central London, or he can send the money from his nearest agent. Mohamoud prefers to send money through his nearest agent, who is quite near to where he lives and is also a personal acquaintance through their respective families. The agent's shop is called Khayraad, and is managed by Adani's friend, Mohamed Bashe. He gives the money he wants to send to his family to the agent, plus the commission, which is 5 per cent or less depending on the amount. Mohamed Bashe completes the necessary documentation and records all the details of where the money is to go. He records the transaction, gives Adani the payment voucher and confirms that the money has been sent. Adani's relatives in Somaliland receive an SMS message almost at once, saying that money is waiting for them in their nearest Dahabshiil office.

Remittances are a vital source of money for thousands of very poor people in Somalia and Somaliland. They are also an important source of capital for local business people, as well as supplementing what are often very low incomes they earn from their businesses. Twenty small business owners in Hargeisa were asked about the role played by remittances in their businesses. Fifteen of the owners were men and five were women, and their average monthly turnover was around $4,000. They were not very small 'subsistence' businesses, but typical small urban manufacturers, employing an average of four workers. Their answers to the PPI survey showed that the likelihood of their being below the $1.50 a day poverty line was well below 20 per cent. Eighty per cent had taken loans from their family and friends over the past three years, and 30 per cent had borrowed from remittance companies.

All the respondents said that their businesses contributed more than 75 per cent of their household income, but half of them said that they also receive remittances from abroad to supplement their incomes. Nearly all of the respondents' children were at school, but several remarked that this was only possible because of the remittances they received.

Many of the business owners said that their livelihoods were much less vulnerable because of their remittances. Mhd Tamir, a mason, said:

> The money we receive from our relatives substantially improves our livelihood, we are much better off than we were. I work as a mason, but the job is irregular, so the money we receive from my brother-in-law in the UK is vital for our livelihood. That is how we can pay school fees for the children and for the food that we eat.

Amal Abdi, a client of Dahabshiil in Hargeisa, said, 'Without the money sent by my brother, my life would have been different. My husband is currently unemployed and the money we receive every month really supports us'. Hassan, father of five children in Hargeisa, said:

> The remittances we receive are important for our livelihood, we are grateful to my sister and her husband who help us pay rent and school fees. We saved a small amount from the remittances and have started a small street business selling clothes in Hargeisa. My sales are improving and I hope to cover all our expenses in the near future.

Most people believe that the remittance system will continue to serve the Somali people. Advanced communication technology accelerates payments and provides real-time information to senders and recipients, informing the clients about the status of their remittances. Dahabshiil itself continues to diversify its services to other sectors such as telecommunication, which complements its money transfer and banking business.

Dahabshiil plans to establish a mobile banking system called E-Dahab that will allow customers to transfer money on their mobile phones. This will enable their customers to transfer money more quickly and at lower cost. Dahabshiil has also opened a bank in Djibouti, with a branch in Hargeisa to bring its services to Somaliland.

In spite of many problems, Dahabshill's mission continues to be to strengthen its market position as the premier regional money transfer organization. The company plans further to expand its network of agents throughout the world by building strong partnerships and adding new products and services to meet the growing expectations of its customers worldwide.

References

Dahabshiil.com (2014) 'Dahabshiil wins injunction against Barclays', online at <www.dahabshiil.co.uk/news/2013/11/dahabshiil-wins-injunction-against-barclays.html> accessed 10 April 2014.

Government of Somaliland (2012) *Somaliland National Development Plan 2012 to 2016* Hargeisa: Government of Somaliland.

Maimbo, S. (Ed.) (2006) *Remittances and Economic Development in Somalia: An Overview*, World Bank Social Development Papers/Conflict Prevention and Reconstruction, No. 38, Washington DC: World Bank.

Shire, Saad A. (2004) 'Transactions with homeland: Remittance', *Bildhaan: An International Journal of Somali Studies* 4, Article 10, page 92-103.

About the author

Abdi Abokor Yusuf is a Programme Specialist for the United Nations Development Programme (UNDP), based in Somaliland. He has an MBA from Preston University, Islamabad Campus and has worked for the UN in Somalia in different capacities for over ten years.

PART TWO
Commodity foods

CHAPTER 8

Nyirefami millet – a traditional Tanzanian crop, marketed in a modern way

Jimmy Ebong and Henri van der Land

Abstract

Nyirefami Limited is a family business that processes and markets millet, sorghum, maize and banana flour. At least 400 farmers are linked to Nyirefami. This case study documents the inclusive value chain initiated by Nyirefami Limited. Data were collected through key informant interviews with representatives of various actors along the chain, focus group discussions with farmers and a sample survey of 20 smallholder farmers supplying millet to Nyirefami. Thanks to participating in the millet value chain, it is probable that 55 per cent of the interviewed farmers increased their income above the poverty line of US$1.25 per day.

Keywords: millet; Tanzania; smallholder farmers

The millet business

Finger millet has for a long time been an important crop for the people of East Africa, including Tanzania. Finger millet flour and yeast have been used to make traditional local beers such as *mbege*. Such local beers have remained important in the cultures of many tribes in East Africa. Finger millet also has industrial uses, extending beyond the traditional production areas. It can be used as livestock feed, for industrial brewing and in a variety of food products.

Over the last decades, the demand for processed foods has been increasing, especially in urban areas. The rural millet markets still mainly focus on beer production. However, there is an increasing demand from a growing urban population for conveniently and hygienically packed local foods, including millet and millet products. This has led to an increase in demand for Nyirefami's products. This case study is about the millet value chain; it presents how Nyirefami became the market leader for industrially processed millet flour and how it has developed a value chain that includes a large number of smallholder farmers.

Nyirefami Limited is a privately owned grain flour processing company situated in Arusha, Tanzania. In the 1970s the owners, Mr and Mrs Nyirenda, began repacking various food products such as salt, spices, rice and powdered milk in their backyard and sold these at the local market. Packing milk powder was their most lucrative business, until they were forced to stop because as an informal business they could not obtain a dairy processing licence, and they shifted to processing local staple foods such as maize, finger millet and sorghum. Business increased and the company was formally registered in 1981.

Nyirefami is managed by Mr and Mrs Nyirenda and today the company is the market leader in processed millet products, supplying around 15–20 per cent of the national market. Millet flour accounts for 75 per cent of the company's turnover. The company sells other types of flours too: maize, sorghum, wheat and banana. Grain flour is a staple food for 75 per cent of the population in Tanzania, with a similar figure for the neighbouring countries, so the market is huge. Nyirefami's strength and its competitive advantage lies in selling only high-quality products. There is competition from local companies, for example Jamahedo, Afri Youth Products, Mama Lishe, Delush and Lam Products, but Nyirefami has managed to take advantage of its early entry in the market and is able to maintain its position due to consistent quality, affordable prices and good presentation of its products.

In the beginning, the growth of Nyirefami was slow, but its owners persisted in their belief the business was viable. They found creative ways to respond to challenges. For instance, they drilled a borehole for the factory to tackle the erratic water supply from the municipal water line, which hindered production. This appeared to be a very good investment because of price increases for the water supplied by the municipality. Responding to quality problems in the millet supplied by farmers, the company bought a threshing machine for farmers. This enabled them to increase greatly the quality of their produce. To overcome the limited space and lack of storage at the company's premises in Arusha, 3 acres of land were purchased at Kisongo, where a new factory will be constructed. Nyirefami's operations were slowed down by stones and other foreign material present in the millet delivered to the factory. In 2007 Nyirefami managed to apply for and obtain a grant of $250,000 from the United States African Development Foundation (USADF) to buy a de-stoner. Although Nyirefami would have grown without this assistance, this grant helped to further improve its processing line and to speed up its development.

Over recent years the company has experienced impressive growth in its sales, revenues and turnover. The company's turnover was $203 million in 2011 and $219 million in 2012, an increase of 8 per cent. Nyirefami's income before tax was TSh26.9 billion ($16.9 million) in 2010, TSh42.2 billion ($26.4 million) in 2011 and TSh48 billion ($30 million) in 2012.

Nyrifami's turnover and profits are impressive enough to attract commercial banks. However, the company is not interested in borrowing from conventional commercial banks because they charge high interest rates. The company prefers to borrow from Root Capital, which is less expensive. Since 2011, Nyirefami has obtained three loans from Root Capital. The first two were for TSh150 million ($93,800) each and the last one was for TSh250 million ($156,000). The interest rate for the first loan was 14 per cent, the second 16 per cent and the last 18 per cent, and all the loans were in local currency. The first and the last loans were for one year, the second for 18 months, and they all had a grace period of three months. The company managed to pay back the second loan in six months. The loans were mainly used for working capital, but they also borrowed TSh12 million from Equity to acquire an automatic packing machine. The loan will be paid back in two years, and payments will only start when the machine has been installed. The interest rate for this loan is 45 per cent.

Nyirefami Limited sources its millet mainly from Singida, Manyara, Dodoma, Arusha, Kilimanjaro and Sumbawanga, located in northern Tanzania; buying much less from the central and southwestern parts of Tanzania. Traders from as far as Zambia have sometimes also brought millet to the company. Millet flour is sold by Nyirefami to distributors in Arusha, Moshi, Dar es Salaam, Tanga and Mwanza. Their main markets are the cities on the coast, reaching out to Mwanza at Lake Victoria. However, Nyirefami's millet flour can also be found in smaller towns. Occasionally, Nyirefami sends its millet flour to the UK to serve migrant families from East Africa.

In spite of the importance of millet to the livelihoods of millions of smallholder farmers (especially in semi-arid areas), and its nutritional value and processing potential, finger millet has received little attention from either the public or the private sector. Local demand for processed millet products has remained low and consumers prefer to process millet themselves, in small quantities, at their homes. Other foods crops such as banana, sweet potato, maize or cassava are less difficult to process, providing an ideal alternative to millet. Government policy prioritized maize as a national food security crop, paying little attention to millet.

The recent developments in millet production and processing are mainly private-sector driven. The main driving force is an increased demand for processed millet products, especially from the growing urban population, to which the private sector has responded very adequately. Tanzania's food processing industry is dominated by beer breweries, juice factories and wheat and maize flour mills, but millet processing is on the rise.

The millet value chain

Production

At least 400 smallholder farmers are linked to the Nyirefami value chain: about 200 in Hanang district, 100 in Kondoa and 100 in Singida. Occasionally, traders from other areas such as Sumbawanga buy millet from farmers to sell to Nyirefami.

A threshing machine takes about half an hour to produce 100 kilos of clean millet seed of good quality. Without the machine, it takes a full day's work and the quality is lower. Some farmers have been able to buy threshing machines, but those who do not have their own machines supply lower quality millet and get a lower price. To guarantee a consistent supply of high-quality millet, Nyirefami started contract farming where it supplied seeds and threshing equipment to farmers. The farmers in return would produce millet and sell it to Nyirefami. Most millet farmers save seed from their own harvest for the next season, leading to poor yields. The improved seeds supplied by Nyirefami substantially increased yields. Threshing improved the quality of the millet and as a result the farmers received a higher price. However, contract farming did not work since a number of farmers sold the millet that they had produced from Nyirefami's seeds to other buyers. Disappointed with the result, Nyrefami stopped all contract farming arrangements. It no longer supplies seeds to farmers and the threshing machine is now operated by an agent who charges the farmers a commission for threshing their millet.

Trade

The number of traders fluctuates every year but in 2013 about 14 traders supplied millet to Nyirefami. The company has had a long relationship, for more than 20 years, with an agent in Singida. This agent has a number of middlemen who buy millet for him. He also manages a millet threshing machine for Nyirefami. Nyirefami regularly gives advances to its suppliers, especially to those with whom it has worked for a long time.

Processing

The production process is quite simple, involving semi-mechanized and manual operations. Millet is washed and dried manually, which slows down the production process. Before the millet is milled, the company's laboratory tests it for *E. Coli* bacteria. The company has the capacity to mill 8 tonnes per day during a shift of eight hours. It has a maximum capacity of 1,920 tonnes a year. At the moment the company is only milling 500 tonnes of millet per year, utilizing only 26 per cent of capacity. This is mainly because of its limited capacity to wash, dry, hull, de-stone and package. The factory has three grain milling machines, one gyro machine, one de-stoner, one cereal dryer and one automated packing machine.

Wholesale and distribution

Distributors collect the millet flour directly from the factory. Currently, the company works with nine distributors, covering Arusha, Dar es Salaam, Mwanza, Moshi and Tanga. Some distributors pay upfront to Nyirefami to secure the supply they need, easing somewhat the working capital requirements for the company.

Retail and consumption

The distributors sell to a network of retailers across the country. It is estimated that more than ten retailers source millet flour from each of the nine distributors. Some of the distributors also have their own retail shops. Most of the consumers that buy Nyirefami products are urban and belong to the middle-income category. They appreciate the high-quality, conveniently packed processed product that is produced in Tanzania.

Secondary actors

Extension services to millet farmers are limited since millet is among the least prioritized crops in the national agricultural policy. Agricultural research in millet is limited too, with the exemption of the ICRISAT-HOPE (International Crops Research Institute for the Semi-Arid Tropics and Harnessing Opportunities for Productivity Enhancement) project, which chose millet as one of its priority crops. With the assistance of DFID and the Department of Research and Development (DRD), this project has been able to release two new millet varieties in Tanzania. These varieties have higher yields, mature early, are resistant to blast and more tolerant to drought compared with traditional varieties.

The value chain map

Table 8.1 below presents costs, profits and gross margins along the millet value chain. Costs are calculated for 1 kilo of millet, and the average prices in 2013 were used. The information was obtained from various actors in the value chain; but only direct costs could be collected.

Costs and selling prices increase along the value chain. Distributors and retailers have the lowest profit margins per kilo but it must be recognized that they handle large volumes and therefore obtain a good income from the millet trade. Farmers make the highest margins per kilo of millet grains.

Table 8.1 Millet value chain: costs, revenues and margins at different levels in the chain

	Farmer	Trader	Nyirefami	Distributor	Retailer
Buying price	0.00	0.46	0.56	1.30	1.42
Direct cost ($/kg)	0.15	0.07	0.16	0.06	0.03
Total cost ($/kg)	0.15	0.53	0.72	1.36	1.45
Selling price ($/kg)	0.46	0.56	1.30	1.42	1.54
Margin ($/kg)	0.31	0.03	0.58	0.06	0.09
Gross margins (%)	67	5	44	4	6
Value added ($/kg)	0.46	0.10	0.74	0.12	0.12
Value share of end market price (%)	30	6	48	8	8

Source: Field data

86 COMMERCIAL AND INCLUSIVE VALUE CHAINS

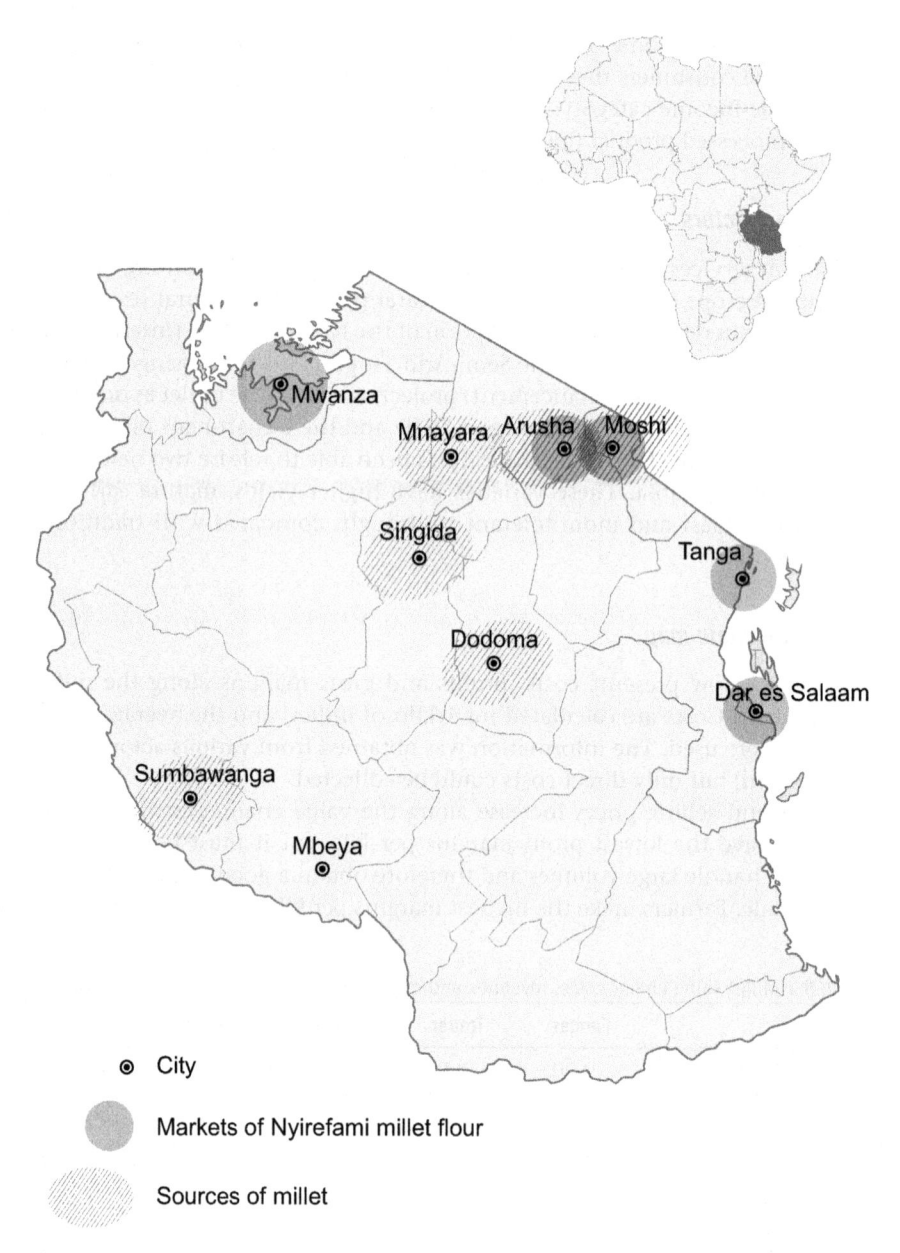

Figure 8.1 Map of Tanzania showing Nyirefami's main sources of millet and end markets of millet flour

Why this value chain is inclusive

A survey was done among 20 smallholder farmers selling directly or via traders to Nyirefami using the Progress out of Poverty Index (PPI) instrument. Before joining the value chain, 17 (85 per cent) of the respondents were considered poor, living on less than $1.25 per day, while 3 (15 per cent) were relatively well off. After joining the value chain, 6 (30 per cent) of the respondents were considered poor, living from less than $1.25/day while the remaining 14 (70 per cent) of the respondents were above the $1.25/day threshold. For this small group of farmers, taking part in the Nyirefami value chain improved the livelihood of 11 (55 per cent) of the respondents. It must be noted that due to resource limitations, the sampling was done in a convenient and not a scientifically robust way. In addition, the number of respondents was too limited to justify firm conclusions on the inclusiveness of the value chain.

Besides the direct benefits to farmers linked to Nyirefami, the company provides employment to 28 people: 9 men and 19 women; 20 are full time and 8 are part time. Their salaries range between TZS1,200,000 ($750) and TZS150,000 ($94), while the average salary is TZS675,000 ($422).

The company contributes to the national pension fund (National Social Security Fund) for all permanent staff and has an education facility for the children of their employees. The company is also paying school fees for about 70 orphans. Besides farmers and employees, there are other actors along the chain such as traders, distributors and retailers, who benefit from Nyirefami.

Box 8.1 Testimony of trader with Nyirefami Millet

Mr Sima Mwangi has a store where he keeps stock of a range of commodities. He has supplied 40 tonnes of millet to Nyirefami. He earns a gross profit of around $200 per tonne. He supplies millet four times a year (each time at a volume of 4 tonnes), leaving him with an income of $3,200. His income could grow further if he managed to increase the volume he sold to Nyirefami. About 40 per cent of his business is sunflower, while millet is second, with 30 per cent of turnover. But he prefers millet since it gives him a higher margin. He is grateful that Nyirefami has given a boost to the millet trade.

About the authors

Jimmy Ebong is a private sector development consultant who routinely conducts research and offers advisory services to businesses in Eastern and Southern Africa.

Henri van der Land is managing partner of Match Maker Group, a consultancy firm that conducts value chain studies, provides business development services and manages an investment fund targeting small businesses in East Africa.

CHAPTER 9
Rice – smallholder farmers in Malawi can be profitably included

Rollins Chitika

Abstract

The majority of farmers in Africa are small scale and play a vital role in food production. As technology improves, agricultural production is changing and small-scale farmers and businesses are in danger of losing their means of living. This provokes a number of questions: can small-scale producers, traders and retailers be part of this change? How will their inclusion or exclusion contribute to the long-term profitability of the entire agricultural sector? This case study from Malawi shows that a value chain can be inclusive and profitable to all parties.

Mtalimanja Holdings Limited (MHL) is a privately owned company that raised an investment capital of US$5 million. This case study explores how inclusive a commercially oriented value chain can be and how the benefits accrue to the different actors in the chain. The MHL value chain shows that small-scale rice producers are essential for their business model. The initiative is likely to boost production in the long run and improve the lives of thousands of small-scale farmers.

Keywords: rice; Malawi; smallholder farmers; outgrowers

Background

In the Malawian economy, agriculture is the most important sector. It contributes more than 35 per cent to the gross domestic product, it employs about 85 per cent of the workforce and contributes about 70 per cent of the country's foreign exchange earnings. Over 80 per cent of households in Malawi are engaged in agriculture. Most of them operate at a small scale, with an average landholding size of less than a quarter hectare (Government of Malawi, 2009). The major cash crops are exported, including tobacco, tea, cotton and sugar. Tobacco alone accounts for more than 50 per cent of the country's export earnings.

Rice is also one of the country's key staple crops and has been promoted by the government, pursuing diversification of food production and exports. It was identified as one of the main cereals, along with millet and sorghum, under the government's Agriculture Sector Wide Approach (ASWAp), which is Malawi's agricultural development agenda as proposed and implemented by

http://dx.doi.org/10.3362/9781780448671.009

the Ministry of Agriculture and Food Security. Two rice varieties are commonly produced in Malawi: Kilombero and Faya. Both are long-grain, aromatic types of rice. The crop is mainly grown in low-altitude areas along the shores of Lake Malawi and Lake Chirwa and along the banks of the Shire River. As a rain-fed crop it is generally planted in December and harvested in May/June. When grown as a winter crop it is planted in July and harvested in November. As the majority of producers grow rice as a rain-fed crop, prices slump in June/July.

Rice is a staple food in Malawi. The markets for rice are differentiated, ranging from expensive premium-quality rice to cheap broken rice. The by-products of rice milling, the husk and bran, are used as animal feed or for making briquettes. Currently Malawi imports over 240,000 tonnes of rice at a cost of over $3.8 million a year. Imports are mainly from the United States, India and South Africa. In Malawi rice only covers 3 per cent of the land suitable for food crops. The production volume is not more than 3 per cent of total food production. The importance of rice is small compared to maize, the country's main staple food. Nevertheless, Malawi has the potential to expand domestic rice production thereby decreasing the need to import from the major rice producing countries.

Mtalimanja Holdings Limited (MHL) is a privately owned company founded in 2012. It processes rice and other crops for the national market, but it also has the ambition to start exporting. MHL's rice processing plant is located in Nkhotakota district, on the shore of Lake Malawi. The company owns a 70-hectare irrigation scheme located near its factory.

MHL purchased its rice-milling machinery from China. The supplier provided technical assistance with the equipment. The company invested about US$5 million in the business, which was nearly all financed by shareholder capital. Attempts to access loans from commercial banks failed, as they did not want to offer loans to start-ups. The company has not received any grants, loans or equity from donors. The Chinese provided an extension expert who advises rice farmers on good agricultural practices (GAPs) in rice production. MHL does not pay the expert but provides him with accommodation and a modest living allowance.

MHL plans to procure rice from 7,000 small-scale producers, mainly located in the Nkhotakota district. MHL intends to process 200 tonnes of rice per day. To achieve this, it needs to boost rice production in Nkhotakota and other rice growing districts. It does this by providing the farmers with a guaranteed market.

MHL has set up its value chain in such a way that the poorer chain actors will directly benefit. As a big commercial company MHL could also have opted for sourcing from large-scale farms, and thereby excluding the small-scale producers and traders. Instead MHL organized its business plan in such a way that it is commercially viable but also inclusive, benefiting poor chain players in the long run.

Figure 9.1 shows the various ways by which the value chain links production to consumption. The emphasis is on how the different actors are benefiting

Major Activities

Input suppliers (seed from Lifuwu Research Station, fertilizers from ATC, ARL, Farmers' World and Rab Processors)

Benefits to Input Suppliers: MHL is creating a demand for inputs

Small-Scale Farmers
7,000 farmers from Nkhotakota district provided with inputs
Other Farmers
May also sell to MHL

Benefits to Small producers:
- *Land*
- *Soft-loan*
- *Extension services*
- *Ready output market*
- *Mechanized farming*
- *Access to financial services*

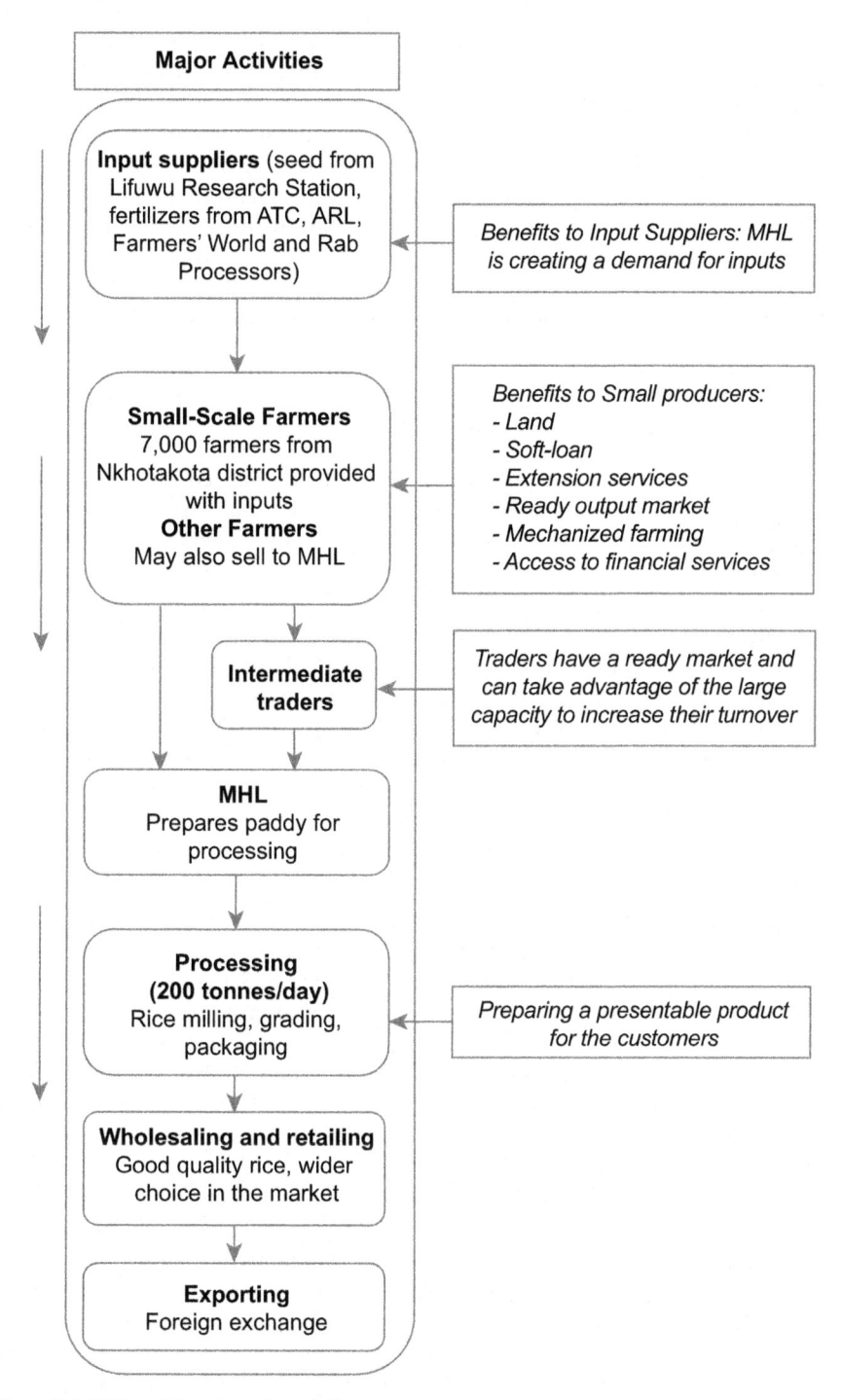

Intermediate traders

Traders have a ready market and can take advantage of the large capacity to increase their turnover

MHL
Prepares paddy for processing

Processing (200 tonnes/day)
Rice milling, grading, packaging

Preparing a presentable product for the customers

Wholesaling and retailing
Good quality rice, wider choice in the market

Exporting
Foreign exchange

Figure 9.1 MHL and the rice value chain

from being part of the value chain, and it demonstrates that a big company such as MHL, with a large commercial investment, can include small-scale actors in a value chain.

From MHL's perspective, its business must generate benefits for all actors in the chain, actively including the poor but also creating opportunities for more affluent actors. MHL is building an inclusive value chain that is commercially oriented, that promotes private-sector participation and also provides benefits to the wider economy. So far MHL has reached about 1,200 farmers in Nkhotakota, although not all have yet been able to sell to the company. The goal is to source from 7,000 farmers within the next 3 to 4 years. In the next sections the different value chain actors are introduced.

Input suppliers

MHL buys fertilizer and certified rice seed from specialized input companies. Seed and fertilizer cover about 20 per cent of the total variable cost for rice production. Fertilizer prices are volatile since it is imported and prices depend on changes in the world market. MHL stimulates the demand for inputs by providing these to farmers. When farmers sell their rice to MHL, the cost of the inputs is subtracted from the money they get for their rice. The company runs demonstrations with farmers showing how important the inputs are for making extra money. With the increased interest in using inputs, more farmers will be willing to buy them from the company. Unlike individual farmers, MHL can buy the inputs in bulk, which means it can negotiate a lower price, to the benefit of the farmers.

Small-scale farmers

Small-scale farmers benefit from this commercial value chain in a number of ways. At the moment around 220 farmers are cultivating rice on the 70 hectares of irrigated land under the Bua Irrigation Scheme, which is owned by MHL. Thanks to irrigation, farmers are able to grow rice twice a year on these plots. In the 2012–2013 season, each farmer cultivated rice on an average of 0.4 hectares, which produced around 24 bags or 1,200 kilos of rice. In total 285 tonnes of rice were produced and sold to MHL. The company could have opted to grow the rice themselves, but decided instead to involve small-scale farmers.

Thanks to MHL, the farmers have access to seed and fertilizers. The costs of these are paid back, without interest, at the time when they sell their rice to MHL. These inputs greatly improve farmers' productivity.

MHL owns power tillers and tractors that are rented out to farmers to do ploughing and levelling. For this service, farmers pay about $6 per tenth of a hectare. This makes land preparation highly efficient and improves the yield, which is good news for the farmer but also for MHL. Ernest Chipala from the Bua Irrigation Scheme praised the benefits of using a power tiller since it makes farming more efficient and easier:

> In the previous season clearing and levelling were done quickly and easily because we used a power tiller from MHL. Its cost was the same as hiring *ganyu* (manual labour). The cost was MK2,000 or about US$5 per one-tenth hectare plot but later on it went up to MK2,500 because of rising fuel prices. Personally, this service helped me because with the same money that I normally pay a labourer, I hired a power tiller which did the same job quickly and also very well. Land preparation is also hard work, we need more power tillers to do this for us.

The rice farmers are supported by an extension agent from China. MHL has been able to get support from the Chinese government to finance this. The goal is to increase rice yields from the current 2.5 metric tonnes to 6.5 tonnes per hectare. The extension agent provides free advice to farmers on matters such as seed quality, appropriate technology, including the principles of SRI (System of Rice Intensification) and GAPs, which are related to sowing, transplanting, recommended seed and fertilizer rates, pest and disease control.

With the increased yields, a result of the inputs used, optimal land preparation and the technical advice, most producers have been able to double their household income as compared to the previous growing season. Using quality inputs, following GAPs and having a reliable market that offers a good price, has made a big difference for them. MHL paid Mk140 per kilo of paddy rice, while most buyers offered only MK100 per kilo. Two farmers clearly analysed the improvement as follows:

> I started growing rice in Mpamantha Scheme in 2007. We used to work with an NGO called Total Land Care (TLC) until the scheme was bought by MHL. One of the benefits I have seen in the 2012–13 season is that my household income has increased. I used to produce 10–12 bags of rice [600 kilos] on 1 acre. But in the last season I have managed to get 24 bags [1,200 kilos] from the same piece of land. I got MK156,000 [$390] compared to about MK60,000 [$150] that I used to get (Aliyana Sadalaki from Nkhongo Village, Nkhotakota).

> I started growing rice in the scheme in 2001. I cultivate 1 acre and used to get 20 bags [1,000 kilos] or less. In the 2012–13 season I produced 46 bags [2,300 kilos]. My income increased from about MK100,000 [$250] to about MK340,000 [$850]. And I was paid on the spot, I did not need to wait for my money. I bought a bicycle and also paid for my driving licence. The money also helped to cover school expenditures of my two children (Samuel Kamwendo from Kalusa Village, Nkhotakota).

The positive impact of MHL results from its abilty to source the rice it needs mainly by offering farmers an accessible and stable market. The company definitely needs more supply in order to reach the 200 tonnes per day it requires to fully make use of its milling equipment. The fact that MHL is not itself involved in producing rice enables many existing and new players to do business and to profit from the rice value chain. This includes traders: village traders assemble the product and sell it to rural traders who sell it to larger traders who in turn transport and sell it to MHL.

Input suppliers benefit from increased sales of fertilizers and seed. Since MHL is strongly promoting the application of inputs in rice production, the demand for seeds and fertilizer is likely to increase, either directly from MHL or from the rice farmers themselves. As MHL expands, this will mean more business for the suppliers of inputs.

New initiatives are developing thanks to the by-products of rice processing. Husks are being processed into briquettes, while rice bran is sold as livestock feed. Bran in particular is expected to become an important additional source of income for MHL, strengthening its business model.

MHL sells the processed rice within Malawi, a market that is undersupplied and has ample room for expansion. Consumers benefit from MHL entering the market, since it is increasing and diversifying supply. There is also demand for Malawi rice from neighbouring countries, which gives MHL additional options to sell its product.

Financial institutions and the economy in general will benefit from the increase in farmers' incomes and additional business activity by input suppliers and traders. The Malawi Savings Bank has opened a branch in the Bua Irrigation Scheme, responding to the demand for savings and loan facilities from the new rice farmers.

The Malawian government will also benefit from the initiative in a number of ways. The country will save foreign exchange by substituting rice imports with domestic production, and may in the future generate foreign exchange from the rice exports. The increase in rice production benefits producers, traders, input suppliers and consumers alike. The additional income and employment generated in the rural areas stimulates local economic development, contributing to a growing and more dynamic rural economy.

Impact on poverty

At the time of writing it was not possible to measure the progress out of poverty from a sample of rice farmers from Nkhotakota district, since they joined the value chain only recently. There was no baseline data or control group, so no firm conclusions could be drawn. Information on the Progress out of Poverty Index (PPI) was only obtained from 20 farmers who had just started working with MHL, and the results suggest that there was no difference in their poverty situation before and after joining MHL. This result was not surprising considering that MHL had just started its operations. MHL is a long-term investment for its owners and the actors in the value chain, and the benefits are yet to be seen. Nevertheless, the potential to decrease poverty seems to be there.

By 2013 MHL had already invested over $5 million in the rice value chain during the previous two years, and they planned to invest as much as an additional $17.5 million to develop their business. The MHL business model centres around sourcing from small-scale farmers. The producers benefit from the initiative by having access to irrigated land, inputs such as fertilizer and seeds, farm equipment, technical assistance and a secured market. The increase in rice production leads to higher incomes for farmers. Meanwhile input suppliers expand their businesses, traders handle larger volumes of rice, and foreign exchange is saved by import substitution.

Reference

Government of Malawi (2009) *Agriculture Sector Wide Approach (ASWAp)*, Lilongwe, Malawi: Ministry of Agriculture and Food Security.

About the author

Rollins Chitika is an agricultural economist. He is a value chain specialist at the African Institute of Corporate Citizenship (AICC) in Malawi. He researches and develops agricultural value chains, where rice is his main priority.

CHAPTER 10
Angkor Rice – 50,000 Cambodian farmers growing for export

Rajeev Roy

Abstract

Ankor Rice pioneered contract farming in Cambodia, drawing on experiences of contract farming organizations elsewhere. An important strategy was establishing farmer associations within the community, under the umbrella of the Angkor Agriculture Association, which help with credit, supply and technical support. The total number of farmers contracted has grown beyond 50,000, and is expected to continue to grow through word-of-mouth publicity and the company's own increased capacity.

Keywords: rice; Cambodia; Angkor Rice; Neang Malis; contract farming

Background

Cambodia has had a disturbed history since it gained independence in 1953. The Vietnam War spilled into Cambodia, giving rise to the Khmer Rouge who came to power in 1975 and carried out the Cambodian genocide, killing roughly a quarter of the population. After much political turmoil, a coup in 1997 placed power with Hun Sen and the Cambodian People's Party, which still remained in power in 2013.

This period of relative political stability has been good for the economy. Cambodia's government is seen as communist free market with authoritarian leanings. In 2011, per capita income was US$1,040. Even though this is low for the region, it has increased tremendously over the past few years. Average gross domestic product growth for the period 2001–10 was about 7.7 per cent annually, making it one of the fastest growing economies in the world (IMF, 2013).

Agriculture and related activities remain the mainstay of the economy, even though tourism is growing very fast. Just like the rest of South East Asia, Cambodia's climate is dominated by the monsoons. The country's landscape is characterized by low-lying plains and abundant water sources, largely connected to the Mekong Delta and Tonle Sap Lake. Rice is the main agricultural product and it is also a staple of the Cambodian diet. About 65 per cent of the population relies on rice farming as their primary source of livelihood (Nesbitt, 1996). Until recently, about 90 per cent of the total agricultural land in Cambodia was used for rice cultivation. With an annual production of

http://dx.doi.org/10.3362/9781780448671.010

8.8 million tonnes (2011), Cambodia is the eleventh largest producer of rice in the world (MAFF, 2013).

Rice is primarily an Asian crop, with Asian farmers accounting for over 92 per cent of global production. It is well suited to regions with abundant rainfall and availability of cheap labour, as it needs a lot of water and is labour intensive. According to O'Brien (1999) 86 per cent of the total rice cropping area in Cambodia is rain-fed.

The productivity of rice farming in Cambodia is low compared with neighbouring countries. The yield is estimated at about 2.05 tonnes per hectare, whereas it is as high as 3.29 tonnes per hectare in Laos and 4.8 tonnes per hectare in Vietnam (MAFF, 2013).

Angkor Rice

In 1999 Chieu Heung established a rice mill under the name of Angkor Kasekam Roongroeung Co., Ltd (AKR), in Kandal province, in the middle of the rice-growing region. With the drought of 1998, Cambodia's self-sufficiency in rice production was under threat and Heung thought that a rice mill would be a good investment in the face of growing interest in increasing rice production. He also realized the potential of export markets and started researching possibilities.

There was huge demand for Hom Mali, an aromatic Thai rice, not only in Thailand but also in other Asian markets such as China and Hong Kong. Neang Malis is a Cambodian rice varietal similar to Hom Mali and Heung decided to promote the cultivation of Neang Malis for export. It is relatively non-responsive to chemical fertilizers, so it lends itself well to organic cultivation. The high cost of organic certification deterred Heung so he decided instead to contract farmers to produce non-certified Neang Malis using organic methods, in the belief that it would fetch a higher price in international markets, despite the lack of certification. The company sold its aromatic rice under the brand name Angkor Rice.

Reception to the launch of the contract farming operation was lukewarm and the farmers were slow in coming aboard. In the first season only about 100 farmers joined. This was partly due to lack of trust in a new company with an unproven track record. Also, the company's lack of adequate milling capacity discouraged it from aggressively signing on many farmers. This proved to be a blessing in disguise as the company was able to demonstrate the potential for higher income for its contract farmers and it was also able to learn from its mistakes before expanding to more farmers. They also invested in a new rice mill with a processing capacity of 10 tonnes per hour or over 30 million tonnes a year. The growth of contract farming was very impressive and the number of contracted farmers grew to 27,346 in 2003 and 32,005 in 2004.

Angkor Rice promoted contract farming in four rice growing provinces of southern Cambodia, namely Kandal, Kampong Speu, Kampot and Takeo. Initially over 80 per cent of their farmers were from Kampong Speu, which is where the company's milling facilities are located. After a quick survey, the company discovered that distance from the mill was not adversely affecting the returns of the farmers, so they started marketing their contract farming more aggressively in the other three provinces. Today, the number of total farmers contracted has grown beyond 50,000, though they are still concentrated in Kampong Speu.

As a pioneer of contract farming in Cambodia, Angkor Rice drew heavily upon the experiences of contract farming organizations elsewhere. One of its main strategies was establishing farmer associations based on existing community organizations and bringing these under its guidance. These community associations were formed under the umbrella of the Angkor Agriculture Association.

Each community association is managed by a head, a deputy and the village head. The head and deputy are trained by Angkor Rice to understand the basic technical aspects of farming Neang Malis and of organic farming. Each association oversees the progress of its members and reports to Angkor Rice management. The progress report covers every stage of production from ploughing, water management, and transplantation to harvesting. Any issues related to the production process, such as drought, flood, disease, insects and other significant issues that affect production, are reported to the company by the community associations.

The company channels its policies with the help of the associations and provides extensive services through its agents. Each association also provides basic technical advice to its members, advises them against using chemical fertilizers, and helps them grow alternate crops after the rice has been harvested. The associations are also expanding their role by helping members develop other sources of income, like growing vegetables in their gardens and raising livestock.

The Angkor Rice value chain activities

Angkor Rice is involved in every stage of the rice value chain (see Figure 10.1). Its roles include identifying areas suitable for growing fragrant paddy; making improved seeds and technical advice available to contracted farmers; monitoring, identifying and solving production problems; collecting and purchasing produced rice at the company mill; sorting milled rice and packing it into different types; and exporting rice to international markets, including Hong Kong, Australia and some European countries.

At the beginning of the season, farmers enter into a contract with the company. The contract specifies the credit terms, the amount to be repaid, the minimum guaranteed price and the penalties for defaulting on delivery. Farmers also agree to abide by the company's quality control process. Details of the production process to be followed are not specified in the contract.

Around July, the company distributes Neang Malis seeds on credit and buys back the produce from farmers between October and January, depending on the planting cycle. The farmers are paid after deducting the seed credit from the value of the produce sold. It is the responsibility of the farmer to transport the rice to the company mill.

100 COMMERCIAL AND INCLUSIVE VALUE CHAINS

Inputs and technical advice given by AKR	
Farmer	$0.3 per kg

↓

Milling yields about 72% of unhusked rough rice	
AKR	$0.8

↓

FOB price realized by AKR	
Trader	$1.2

↓

There might be multiple layers between trader and retailer	
Retailer	$1.8

↓

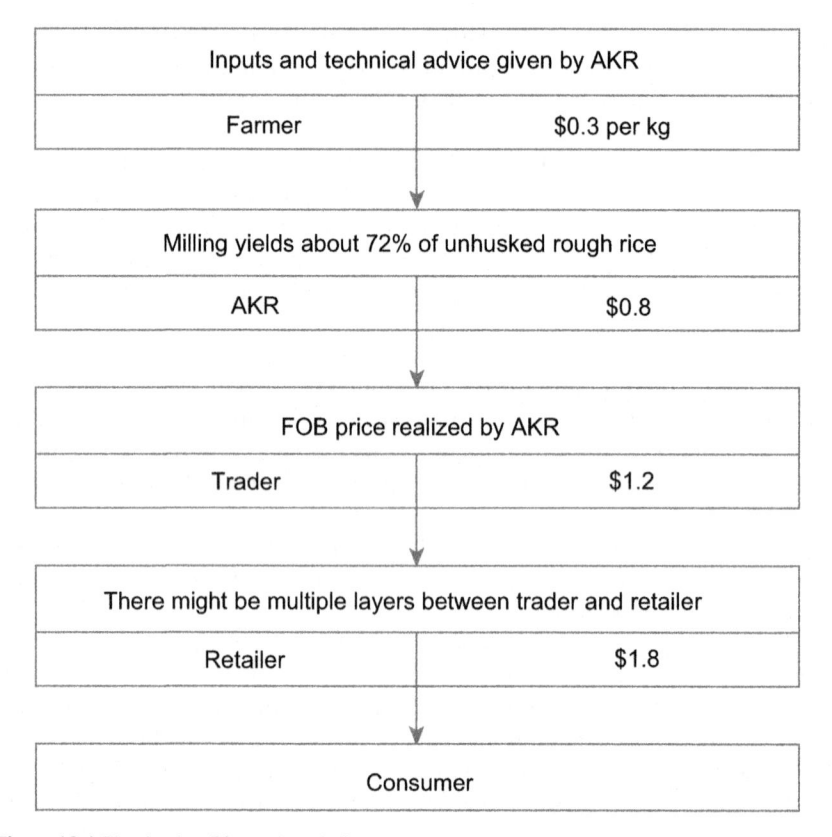

Consumer

Figure 10.1 The Angkor Rice value chain

Farmer benefits

The Progress out of Poverty Index (PPI) survey tool was administered to a small sample of contracted farmers; they answered the ten questions as of the date they were surveyed, and for the time four years ago before they had joined the Angkor Rice value chain. Their answers suggest that their household conditions have improved quite dramatically in the average of four years during which they have been supplying Angkor Rice.

Some of the highlights of the results from the smallholder survey are in Table 10.1.

The improvement in the PPI scores is reflected by increases across all asset scores. So, in spite of a small increase in the total number of family members, which leads to a drop in score, the overall score has increased as families have invested in household and productive assets. The most significant improvements seem to be in house construction materials and in transport. Conversations with the farmers reveal that this increase in investments in household assets was not only due to an increase in income but also due to their view that future cash flows from rice would be more secure than before.

Table 10.1 Results from the survey of smallholders contracted to Angkor Rice

	Before contract	After contract
Number of years the farmer has been in a contract with Angkor Rice		4.1
Number of family members	5.8	6.1
Percentage of children 7 to 15 years attending school	78	82
Percentage of houses made of concrete, stone or cement	61	72
Bicycles per household	0.9	1.2
Motorcycles per household	0.4	0.6
Pumps per household	0.1	0.2
Televisions per household	0.6	0.8

Table 10.2 Economics of a household producing Neang Malis for Angkor Rice

	Cost (CRs)	Revenue (CRs)
Revenue from paddy (including notional cost of paddy for own consumption)		1,800,000
Seed	115,000	
Compost and natural pesticides	230,000	
Rental machinery	95,000	
Irrigation	80,000	
Transportation	115,000	
Other income		600,000
Total		1,765,000 (US$446)

The economics of a household producing Neang Malis for Angkor Rice is approximately as indicated by the figures in Table 10.2.

In these calculations, the cost of labour has been taken as zero as it is usually provided by the family or by the community on the basis of mutual exchange. Also, it had been seen that contracted farmers tend to have larger families. This may be because a larger family means more labour available for rice farming. Also farmers felt that the non-farm income of un-contracted farmers was higher than for contracted farmers. This does indicate that farming Neang Malis is more labour intensive.

The farmers report that the average yield for traditional varietals is better but the Neang Malis fetches a better price. Also, the risk is relatively low because the cash investment is lower in the case of Neang Malis. The company gives seeds on credit and there is less spending on fertilizers and pesticides.

There is a widespread opinion that Angkor Rice is able to teach farmers soil improvement techniques and other agricultural practices that lead to increased productivity in the long run.

The 'graduates'

Several farmers who were previously contracted to Angkor Rice have decided to opt out of contracts and go back to un-contracted rice paddy cultivation, since after a few years of the company's operations, Neang Malis seeds became available in local markets in the regions in which it operates. At the same time, a market for Neang Malis rice started to develop as local traders started purchasing it to sell to Vietnam. Since the un-contracted farmers have an option, they are able to weigh the pros and cons and decide between planting traditional varieties or the Neang Malis seeds. Usually the farmers have found that open market prices were better than the rates offered by Angkor Rice. Also, un-contracted farmers are better able to wait and take advantage of price fluctuations in the market. The farmers who are geographically closer to the market have better access to market information and are able to sell their produce at better prices.

The idea of Angkor Rice was to produce the pure rice without mixing it with any other types. Since the boom in Cambodian Neang Malis, a number of rice traders have been using the Neang Malis name but have packed it with low-quality rice and sold it to foreign rice traders. This practice was destroying the reputation and market for Neang Malis and bringing premiums down. Other companies were not able to enforce the strict standards implemented by Angkor Rice. In May 2012, Angkor Rice secured all rights to use of the name Neang Malis. This has led to an outcry from other traders and farmers but the decision is unlikely to be reversed, at least for a while.

Criticisms

The farmer contracts do not mention any penalties for the company if it fails to buy the rice at the contracted prices. While the contract specifies the minimum purchase price, it does not specify the terms of purchase in detail. Farmers often feel that the company uses technicalities to lower the price or even reject the product after it has been transported to the company mills.

Many farmers feel that the farmer associations are not actually good for the farmers. They feel it is simply a mechanism to recruit farmers and help enforce the contracts. At present, these farmer associations are very tightly controlled by the firm and have little bargaining power to negotiate favourable terms for the farmers.

Another major criticism is that the company contracts farmers who have larger farms and ignores farmers with smaller landholdings. This is borne out by the fact that, according to data from the company, contracted farmers have larger landholdings compared to non-contracted farmers. But company officials say that the difference is not very great and the company has no particular desire to recruit very large farmers. The relatively larger landholdings may reflect the scale requirements for contract farming. Farmers tend to split their land into two parts, land for their own consumption and land for commercial farming. Farmers with smaller holdings have less land left after planting for their own domestic use.

Angkor Rice collected data from existing farmers and concluded that to be efficient farmers need to have at least one hectare of land. So they made this a condition to enter into a contract with the company. The requirement of one hectare may be because of the company's experience that small farmers are more likely to break the contract as the cost of breaching the terms are less. Also, the transaction costs go up with a decrease in size of the average holding. However, on the recommendation of farmer associations, the company sometimes allows some smaller farmers to band together to make up the one hectare minimum requirement.

Another major criticism of the company comes from their competitors and some traders. Even though Angkor Rice claims to sell organic rice, it is not certified and the company does not monitor the farmers closely so a number of contracted farmers may be using chemical fertilizers in an attempt to improve yields.

Angkor Rice in the future

Every year, Cambodia exports about $3 billion worth of its major export product, garments. The government sees rice as having the potential to become the other major export commodity. In 2011 it decided to promote rice paddy production and rice export, aiming to increase Cambodian rice exports to over one million tonnes a year by 2015. Angkor Rice hopes to cash in on this renewed focus on rice. It has made significant investments in its in-house rice research facility and has set up two power generation units based on rice husks.

The company realizes that the key to higher volumes will be higher milling capacity and it is planning to increase their milling capacity by 2014. At the same time, it hopes that good word of mouth from existing contracted farmers will help to increase the number of contracted farmers, especially outside Kampong Speu province.

Box 10.1 Stories from two rice producers selling to Angkor Rice

Kaew Sophalis is 45 years old and lives in Khom Prey Rumduan in Kampong Speu province. He lives with his wife, their two daughters and his mother. His eldest son is at university and comes home during the holidays. He was one of the first to enter into a contract with Angkor Rice and he has certainly not regretted it. The income from growing Neang Malis has been good. He has used the money to buy a motorcycle and a second bicycle for his family. He says, 'Most of our village grows rice for the company. Although the rice can be grown only once a year, the yield and the price are good'.

He is happy with the support he has received from the Angkor Rice company. They have taught him the importance of crop rotation, using compost and organic pesticides.

> He acknowledges that there has been an unprecedented run of good luck with the region not having suffered a drought or disease for a number of years now. The community association head and its deputy have always been there to help him whenever there were problems for instance with the provision of credit or seeds. The company technicians were there to guide and give tips whenever necessary.
>
> Chhan Samphan and Som Phuang live with their three children and Samphan's brother in Khom Prey Rumduan village in Kampong Speu province. The family has sold its rice under contract to Angkor Rice for over 10 of the past 12 years. Generally their life has improved: during this period they bought a TV, a motorcycle and furniture for the house. They do feel that they can do much better and that they might even have to look beyond Angkor Rice to further improve their income.
>
> And after being on contract for nine years, they decided to grow and sell on their own. They saw that the open market prices were almost always better than what Angkor Rice offered. After so many years on contract, there was very little guidance that the company technicians could provide anymore. The couple was very experienced with the process of cultivating rice.
>
> For two years they sold their rice in the open market to traders who sold the product in Vietnam. The price they got was good but highly variable. Then some of the major rice merchants in the market place near their village decided to migrate to another place closer to the Vietnamese border. The new rice merchants were not as professional nor reliable as the previous ones and Samphan decided to go back into contract with Angkor Rice. Though the company is usually reluctant to take back people into their fold, the village headman personally intervened and put in a good word for them.
>
> The couple is now trying to expand their kitchen garden. They have seen a sharp increase in the prices of vegetables in the local market and they intend to cash in on that to supplement their income from growing for Angkor Rice.

References

IMF (International Monetary Fund) (2013) World Economic Outlook Database, online<http://www.imf.org/external/pubs/ft/weo/2013/02/weodata/weorept.aspx> accessed 7 October 2013.

MAFF (Ministry of Agriculture, Forests and Fisheries) (2013) online <www.maff.gov.kh/ >, accessed 7 October 2013, in Khmer.

Nesbitt, H.J. (Ed.) (1996) *Rice Production in Cambodia*, Phnom Penh, Cambodia: University Press.

O'Brien, N. (Ed.) (1999) *Environment: Concepts and issues, a focus on Cambodia*, UNDP/ETAP Reference Guidebook, Phnom Penh: Ministry of Education, Government of Cambodia.

About the author

Rajeev Roy was an entrepreneur and currently teaches entrepreneurship. He has been engaged in several entrepreneurship development initiatives across the world. He is also mentoring several start-ups in India.

CHAPTER 11

Moksha Yug – Indian dairy farmers don't have to be in cooperatives

Chandrakanta Sahoo

Abstract

This case describes the commercial milk value chain that Moksha Yug has developed in Karnataka state of Southern India. Most dairy producers in India work through state-sponsored cooperatives, but Moksha Yug's value chain shows how it is possible and profitable for a private for-profit business to reach out to very small-scale dairy farmers without the intermediation of cooperatives.

Keywords: dairy; milk production; India; Operation Flood

Dairy in India

Operation Flood, a project of the National Dairy Development Board (NDDB), was the world's biggest dairy development programme and it transformed India from a milk-deficient nation to the largest milk producer in the world. India accounted for about 17 per cent of global output in 2010–11. The project doubled the milk available per person and made dairy farming a sustainable rural employment generator.

Operation Flood created a national grid linking milk producers with consumers, ensuring that the producer got a major share of the price consumers paid. Operation Flood was based on the formation of village milk producers' cooperatives, which procured milk and provided producers with inputs and services. The model was extended nationwide after its initial success in a single dairy cooperative in Anand in the state of Gujarat.

During its first phase from 1970–79, Operation Flood linked 18 premier milk-producing areas with consumers in India's four metropolitan cities: Delhi, Kolkata, Mumbai and Chennai. The focus was on increasing both production and procurement.

By the end of the second phase in 1985, 43,000 village cooperatives with 4.25 million milk producers were integrated into a self-sustaining system. Production of milk powder in India increased from 22,000 tonnes before the project started, to 140,000 tonnes by 1989, with all of the increase coming from cooperative dairies set up under Operation Flood (NDDB, 2013).

Operation Flood's Phase III from 1985–96 consolidated India's dairy cooperative movement, with the total number of cooperatives going up to

http://dx.doi.org/10.3362/9781780448671.011

73,000. During this phase, the number of women members and women's dairy cooperative societies increased significantly (NDDB, 2013). According to the Rural Economic and Demographic Survey (DAHAF, 1999), in 1999, 45 per cent of rural Indian households owned at least one cow or one buffalo.

As a result of Operation Flood, the cooperatives were successful in cutting out middlemen, leaving the private sector with a very small role to play in the Indian dairy sector.

India now leads the world in milk production by a wide margin. The United States come second, and is followed by much smaller producers such as China, Pakistan and Russia.

The dairy industry in Karnataka

In Karnataka, the apex dairy cooperative is the Karnataka Cooperative Milk Producers' Federation Limited (KMF). KMF sells its products under the brand name Nandini. It procures about 5 million litres of milk daily and, after Gujarat, it is the largest state milk cooperative in India. Unfortunately, outside KMF's influence, the rest of the dairy sector in the state has not done as well and Karnataka ranks eleventh overall in dairy production in the country.

At the conclusion of the third phase of Operation Flood, the basic infrastructure for milk production in Karnataka had been set up and this encouraged commercial dairy farming on a small and medium scale. The milk cooperatives found their supply rising and slowly they focused on procurement only and their role in livestock development and farmer outreach started to decline.

Table 11.1 Development of milk production in India

Year	Milk production (million tonnes)	Per capita availability (gram/day)
1950–51	17.0	130
1955–56	19.0	132
1960–61	20.0	126
1968–69	21.2	112
1973–74	23.2	110
1979–80	30.4	125
1985–86	44.0	160
1990–91	53.9	176
1995–96	66.2	195
2000–01	80.6	217
2005–06	97.1	241
2010–11	121.8	281

Source: DAHAF, 1999, 2012

Procurement from larger dairy farms was easier and the needs of the smallholders were being ignored. As a result, quality of feed and care provided to cattle in smallholder households started going down. Cattle were left to graze on their own and free-grazing animals received very little additional high-quality feed and nutrients.

Mosha Yug Access

In 2006, Harsha Moily set up Moksha Yug Access (MYA). Previously, Harsha, who belongs to a political family, had worked for about 15 years in telecommunications, for private equity funds and agri-businesses in India, the US and the UK. For the first few years, MYA experimented with different business designs before settling down with its current business model. MYA started out as a microfinance business and then added primary healthcare and dairy farming as additional businesses. Even though microfinance continues as a separate business, the focus is now on rural supply chains primarily serving the dairy sector. MYA currently operates 1,200 milk collection centres spread across 1,100 villages in 5 districts in rural Karnataka. Over 15,000 farmers currently supply milk to MYA. The daily total procurement varies seasonally and it has touched 100,000 litres on some good days.

The approach

According to the Department of Animal Husbandry, Agriculture and Fisheries, India is the world's largest producer of milk. In 2011–12 India produced 127 million metric tonnes of milk which accounted for about 17 per cent of global milk production (DAHAF, 2012). At the same time, productivity of Indian cattle is extremely low. The yield per cow per day can be as high as 40 litres in Israel, the US and Canada, whereas in India it is only 2 to 4 litres. The main reasons for the low yield are the following:

1. Inadequate year-round nutrition for cattle
2. Lack of improved breeds
3. Limited availability of artificial insemination services
4. Limited focus on prevention of cattle diseases; most farmers only get veterinary services when their cow is already sick
5. Lack of hygienic practices while milking the cow

The average figures in Table 11.2 were collected from a village which is currently in the network of KMF. Table 11.2 represents a simple profit and loss account for a single cow. The figures are likely to vary a lot across cows but this provides a rough estimate.

Profit is approximately US$46 per annum. This figure looks even less attractive if the opportunity cost of the hours of labour invested in the upkeep of domestic cattle is included. Also, there is the initial investment of Rs10,000 or more for the cow and some investment in the cowshed. The figures in Table 11.2 are not very different from the findings of other studies.

Table 11.2 Profit and loss account for a dairy cow

Source of revenue	Revenue
Sale of milk	Rs7,825
Value of calf	Rs1,200
Value of dung	Rs1,560
Total	Rs10,585
Source of expenditure	**Expenditure**
Fodder	Rs7,510
Veterinary	Rs140
Insemination	Rs110
Total	Rs7,760
Profit (revenue–expenditure)	Rs2,825

Note: US$1 is Rs61.41 (average exchange rate at the beginning of 2014).

As a result of such poor returns, households had stopped focusing on dairy farming as a significant means of livelihood. Moksha Yug saw an opportunity. If they could demonstrate higher returns to farmers then there could be a way to develop a robust supply chain for milk. There were two important elements to this approach.

First, they had to improve yield per cow and the quality of milk to get higher returns for smallholders. The same farmers, with the same cows, had to produce more and better milk. This could be done by getting better feed and veterinary care to the cattle. And second, they had to make it possible for farmers' milk to be collected from their houses.

The MYA business model

For its first few years, MYA procured milk only to pass it on to other dairy firms. The margins in the B2B (business to business) dairy business were very low. MYA used to make about 3 per cent average net margin on the liquid milk it sold to other firms. In 2013 they decided to launch their own dairy brand, Milk Route. Now, its net margin net margin for its consumer products ranges from 4 to 20 per cent: 4 per cent for the ordinary pasteurized milk pouches and up to 20 per cent for the ultra-high temperature (UHT) milk with long shelf life. Milk Route is a new initiative by MYA and accounts for about 30 per cent of the total milk collected. Most of the milk is still sold to food processing units and other dairy firms. In the future, MYA intends to expand the Milk Route brand by introducing yoghurt, cheese and other dairy products.

In 2013, MYA also started 'The Good Chain': a chain of retail stores selling fresh fruits, vegetables and milk. It is loosely modelled on WholeFoods in the US. With 80 per cent of dairy farmers also growing fruit or vegetables, there are logical synergies in the businesses. Fruit and vegetables are procured by the same chain that procures milk. Currently, fruit and vegetables are less than 20 per cent of MYA's turnover.

Rural infrastructure

In the case of KMF, as with other milk cooperatives, the farmers go to a milk collection centre to deliver the milk. The milk quality is checked, the milk quantity is measured and an entry is made in a logbook. Periodically, usually monthly, the payments are made. Sometimes there are small delays in the payments.

If the income from milk is small, there is little motivation to go to a milk collection centre and to wait in a queue to deliver the milk. Later, there is also a queue while getting paid for the milk. MYA's collection system cuts through all that. The milk is collected from the farmer's house itself.

Milk collection centres are located in the villages and they collect milk from the farmers' doorstep twice a day. The collection centres are manned by MYA representatives who sort and grade the milk collected. All centres are equipped with large milk churns to consolidate milk collections. The representatives are trained to use testing tools to measure the quality and quantity of milk collected. At the time of collection, milk quality is tested with a lactometer and the result is shown to the farmers. Every household in the MYA network has a card in which the daily quantity of milk collected and lactometer readings are noted. Payments are made every two weeks.

These collection centres also serve as the aggregation and transaction point for cattle feed, veterinary services, artificial insemination and other farmer services. MYA rents vehicles from local owners to collect raw milk from collection centres and drop it off at its bulk milk-chilling centres. These milk vans have a capacity of 800 litres and usually make two runs a day. They carry the distinctive black and white MYA design but are not owned by MYA. They are rented on a per kilometre basis.

The rural infrastructure MYA has set up is critical to its business model. In India today, the lack of cold storage and chilling facilities in villages has led to a lot of the rural milk produced losing its quality by the time it reaches the processing plant. MYA has addressed this problem through its bulk milk chilling centres that chill the raw milk, ensuring limited bacterial build-up, resulting in better quality milk for its downstream operations and consequently better prices to dairy farmers. Each bulk chilling facility has a storage capacity of 5,000 litres, which allows it to cater to about 15 to 20 villages. MYA also follows the 'golden hour' principle, which entails collecting and chilling milk within the first hour of milking. In the traditional dairy model, players set up large bulk milk-chilling plants of 30,000–50,000 litres capacity. These bulk milk chillers service a wide area. This means that the milk has

110 COMMERCIAL AND INCLUSIVE VALUE CHAINS

to travel a significant distance from the collection point and is not always chilled within the 'golden hour'. In order to stay within that time period, MYA has opted to set up 5,000-litre capacity milk chillers within a maximum of an hour's travel from its collection centres. Here milk from various milk collection centres is consolidated, stored and chilled at 3 degrees Celsius. A single milk chiller caters to about 15 to 20 villages. For optimum utilization, the location of these chillers is determined based on various parameters,

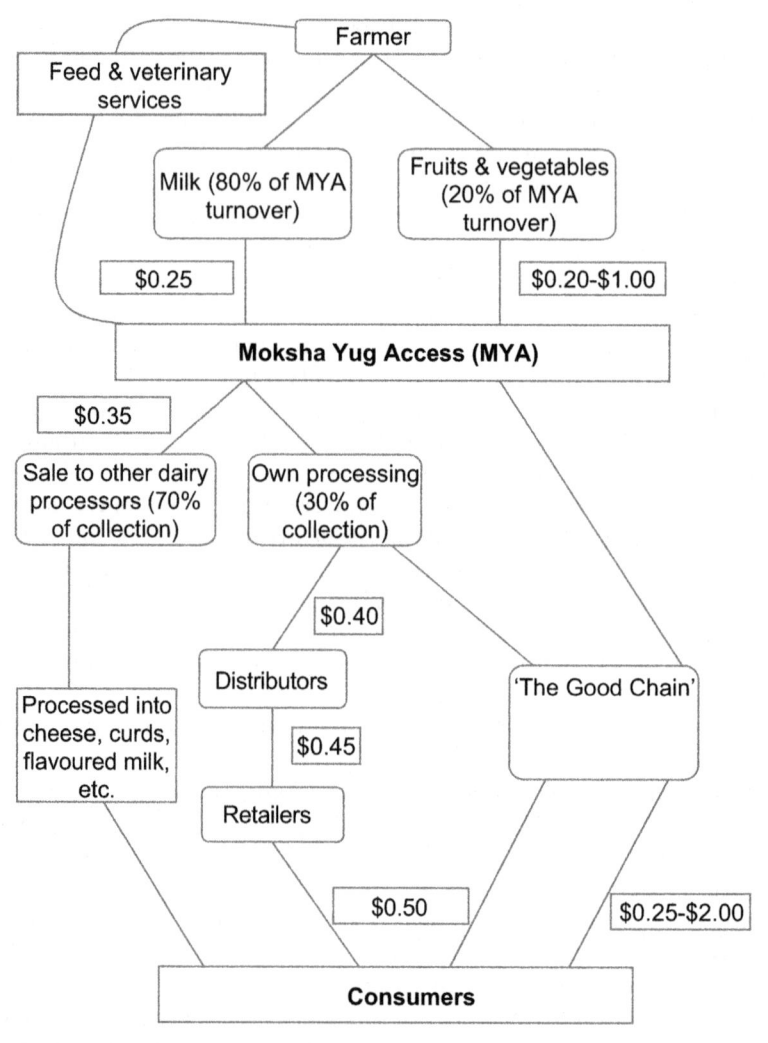

Figure 11.1 Moksha Yug Access value chain
Source: DAHAF, 1999, 2012

including the density of dairy farmers in the area, the amount of milk that can be procured daily, and the road and power infrastructure.

Chilled milk is quickly transported to the processing plants using bulk milk tankers. Currently, MYA has outsourced the processing and packaging of its milk because these activities require heavy capital expenditures – for example, it costs between $550,000 and $900,000 to set up a 100,000-litre processing unit. The current volume of milk procured by MYA does not justify such spending.

Benefit to farmers

The number of farmers in the MYA network has been steadily increasing. Farmers have been reporting that their net income from milk has doubled by following MYA's advice. Convincing the farmers is not an easy task; it takes investment from the farmers before they start seeing the results. Usually the spending on feed can go up by about 30 per cent but the good news is that the farmers are able to see the results quite quickly, usually within a month.

In some cases the higher productivity is not translating into higher milk procurement for MYA as the extra milk is being used for household consumption. Even though MYA would be happier with the additional milk coming to their collection system, they are happy that it is contributing to the nutrition of the families in their network. In the long term, it may translate into purchase of another cow by the family, which will lead to higher collections.

The collection centres are run by locals trained by MYA. These local representatives of MYA act as their brand ambassadors. They are paid a commission on the milk collected and they also receive a margin on the feed sold. So it is in their interest to talk to other villagers and convince them to switch to MYA.

Kategollanahalli is a small village of about 50 households in Tumkur district in Karnataka. Around 35 of the 50 households kept cows but most had only one or two. KMF was already active in the village and all the 35 cow-owning

---

Box 11.1 Three dairy farmers supplying milk to MYA.

Mahadev of Channapanadoddi, Mandya district

Early in his life, Mahadev decided that he would rather be a cattle farmer than a crop farmer. For many years Mahadev supplied milk to KMF and was happy with the price that he received for his produce. However, as local politics changed the way the cooperative was run in his area, his experience soured, and he decided to partner with MYA instead.

Having been a Milk Route producer for over 13 months, he has continually expanded his herd from 8 to 21 cows and supplies 140 litres of milk every day. His wife, Nalina, is also the village milk collection centre supervisor and Mahadev is very supportive of her role: 'I feel happy and proud that my wife is an equal and earning member of the family', he says, 'and by saving income from her supervisory role, Nalina has been able to purchase some gold as well!'.

Mahadev feels that there is a noticeable difference between supplying milk to KMF and to MYA. He points out that the field staff at MYA are more approachable and pleasant to interact with. He also believes the cattle feed is of excellent quality and appreciates its on-time delivery. Operation Milk Rich (MYA's initiative to help farmers improve dairy yields) has benefited him tremendously by providing education on feeding practices. 'Earlier we would just go and dump food in front of the cow, but now we know how much should be fed based on the weight of the cow', he said, noting that this change in practice has also helped him save on veterinary costs.

Mahadev is not without worries and fears that any drop in prices will make it impossible to cover more advanced costs such as artificial insemination and nutrient-rich cattle feeds. If that were to happen, he is afraid that falling prices will force him to sell his cows – a very tough decision for a man who, many years ago, followed his calling to become a dairy farmer.

Varalaxmi of Jyotinhalli, Kolar district

Before Varalaxmi started working in MYA's Milk Route network 15 months ago, she and her husband Naryanappa and their two children lived uncomfortably in a house with a straw thatch roof. The family owned three cows, but struggled to sell the milk to a cooperative that was located more than 1 km from their village. Often, Naryanappa would make the trek to the collection centre, with milk in hand, but the centre would keep unpredictable hours and many times would be closed. Even on the days it was open, Naryanappa felt he wasn't being compensated fairly. It was difficult for the family to count on milk production as a steady source of income.

Since they started selling milk to Milk Route, the family's monthly income has increased by 25 per cent. Varalaxmi and Naryanappa now bring home Rs8,400 for the household each month. Varalaxmi also works as a cook and earns an additional Rs1,000 per month. Together they are earning more for their family than they had previously dreamed of.

Varalaxmi now happily reports that the dairy activity is the household's primary livelihood. MYA's milk collection centre is located in the village, and is much more easily accessible. Varalaxmi says she is treated well by the centre staff. Since switching over to Milk Route, 'cattle rearing has become more profitable for us, and now we would like to buy more cows', Naryanappa says. The family has been steadily saving, and hopes to purchase more cows in the coming months.

With the extra income, Naryanappa fixed up the house: the family now lives in a three-room house with brick walls and a cement roof. Varalaxmi is quick to point out that his house now also has an electricity connection! They regularly buy seeds and fertilizer to grow potatoes, tomato and *ragi* (millet), as other sources of income. The children are now regularly attending school – in the 9th and 11th grades – and have hopes of pursuing higher education.

Lakshamma of Saganhalli, Kolar district

Lakshamma, a 38-year-old widow in Sanganhalli, a village in Kolar district, struggled for years to support her three children on her income from cattle rearing. Since she does not receive a widow's pension from the government, her family depends solely on the cattle for their regular income.

Her daily routine involved trudging 1.5 km each morning to sell milk to KMF milk collection centre (MCC), but she often felt she was short-changed. The MCC frequently weighed the milk inaccurately, especially when the electricity was out. KMF paid her at most Rs16.50 per litre, and payment was often delayed by three to four days. Lakshamma also complained that it was difficult to go to the MCC while taking care of her young children: 'The MCC was far away, and I couldn't leave my kids at home alone. It was especially difficult when it rained'.

A little over a year ago, Lakshamma had the chance to begin selling milk to MYA and she jumped at the opportunity. MYA pays her Rs19.5 per litre, and Lakshamma feels the payments are fair and timely. Lakshamma now feeds her cow and calf the high-quality MYA cattle feed, and she's noticed her cow has increased milk production by 1 to 2 litres a day. The cow produces 8 litres of milk a day, and Lakshamma's income has risen to Rs4,800 per month.

MYA's other services have also helped Lakshamma a great deal. She has taken an advance of Rs3,000–5,000 to help her cover her expenses. She finds the veterinary services particularly useful. 'Earlier there was no vet here. Sometimes government vets

> would come from Bangarpet, which is 20 km away', she explained. Now Lakshamma has consistent and easy access to veterinary services.
>
> Lakshamma and her children currently live in a mud house with a tiled roof, with very few possessions. On 3 acres of land, she grows *ragi* and green fodder and occasionally makes a little extra income if there is any leftover green fodder she can sell. However, working with MYA has allowed Lakshamma to dream big for her children: 13-year-old Bhuvanesh, 7-year-old Kavya and 5-year-old Bhargavi. Lakshamma is determined to give a good education to her children, and they all attend the local government school. When asked what she plans to do with her increased income from MYA, she muses, 'Maybe I'll build a *pucca* [permanent] house and buy gold to give my daughters for their weddings. I'll also make sure my son goes on to higher studies'.

families sold their milk to KMF, which had been operating in the region for several years. The problem was that there was no milk collection centre in their village and they had to go to a neighbouring village to deliver their milk. This was a minor inconvenience but there was no alternative.

MYA opened a centre in the village and trained Anusuya to be in charge of the collection centre. Anusuya went around trying to convince people to give her their milk and a few families shifted from KMF. Over time when others saw the convenience of selling to MYA they started getting interested. Also, they could see that the yield from cows had gone up for the families who joined MYA. In a matter of days, over half the cow-owning families started to supply milk to MYA. Many families are currently selling milk both to MYA and to KMF.

Challenges facing Moksha Yug

MYA has managed to gain traction in a very difficult business. It faces several challenges and there is a long way to go. The biggest challenge is from its competitor. MYA has been able to make inroads where KMF has failed. If KMF improves its performance, it will be more difficult for MYA to make headway with farmers. Also, KMF is an entity that has a number of influential politicians as stakeholders and they may create roadblocks for MYA in the future.

MYA's strategy of reaching underserved and small villages is a good way to get a toehold but scattered and difficult-to-reach milk collection centres will make milk collection comparatively costly and more difficult.

MYA is currently supplying milk to other dairy processors and does not have much of a presence in the consumer market. Being a B2B player, its margins are thinner and there is always the danger of one of its customers backward integrating into procurement. Major international food processing companies including Danone, Nestlé and Brittania have been showing interest in expanding their procurement operations in rural India.

MYA's own brand, Milk Route, is still new and yet to establish itself. The Good Chain is yet another new initiative by the company. While it will go a long way in getting 'The Milk Route' to its customers, this is an entirely new business with challenges of its own.

The company's turnover is approximately $20 million, with over 80 per cent coming from milk and the rest from fruits and vegetables. MYA has already raised about $8 million from Khosla Impact Fund, Unitus Equity Fund and other venture funds. There are plans to raise another $15 million to fund the next phase of expansion. Moily is targeting revenues of over $100 million and wants to expand the MYA network to 100,000 farmers over the next three years.

References

DAHAF (Department of Animal Husbandry, Agriculture and Fisheries) (1999) *Rural Economic and Demographic Survey (REDS)*, Delhi: DAHAF.

DAHAF (2012) *Rural Economic and Demographic Survey (REDS)*, Delhi: DAHA.

NDDB (National Dairy Development Board) (2013) 'Statistics', online <http://www.nddb.org/English/statistics/Pages/Statistics.aspx> accessed 7 September 2013.

About the author

Chandrakanta Sahoo is Assistant Professor in the Department of Management Studies, Madanapalle Institute of Technology & Science (MITS), Andhra Pradesh. He has been teaching in management schools for a number of years and he is interested in corporate social responsibility.

CHAPTER 12

Suguna Poultry – decentralized village production is good business

Malcolm Harper, Rajeev Roy and Phanish Kumar

Abstract

Suguna Poultry is the largest poultry business in India with an annual turnover of nearly half a billion dollars. It provides chicks, feed, training and veterinary support to farmers in return for eggs. This survey of 180 smallholders spread over West Bengal, Odisha, Andhra Pradesh and Tamil Nadu looks at the impact of Suguna's business on smallholders associated with the company. The case gives details of the broad-based positive impact of Suguna's business on producers.

Keywords: poultry; India; broilers; outsourcing

Between 2005 and 2010, chicken consumption in India grew at a rate of about 18 per cent annually, which was almost twice as high as growth in the 1994–2005 period, according to data from the National Sample Survey (MSPI, 2011). The only other food that comes close to chicken in its growth rate is eggs, which grew at an annualized rate of over 10 per cent in 2005–10. The value of poultry meat produced in 2010–11 was over Rs300 billion, or nearly US$5 billion (Seetharaman, 2013).

Farmers in India have moved from rearing country birds to rearing hybrids that ensure faster growth of chicks, increased hatchability, lower mortality rates, higher numbers of eggs per bird, better feed conversion and consequently greater profits for the poultry farmers. Productivity gains of the poultry industry are reflected in relatively low price increases in poultry products over the past five years when compared to other animal protein products. Poultry prices have grown at 12 per cent year-on-year over the period 2008–13 against 21 per cent for overall meat products (MOFPI, 2013), thus providing a cheaper alternative for meeting protein requirements in the Indian diet. Also, there are religious sentiments against the consumption of pork and beef by Muslims and Hindus respectively.

There is significant growth in the Indian poultry industry. About 3 million poultry farmers are employed in the poultry industry, producing poultry and eggs and contributing almost US$6 billion annually to the national income. Additionally 15 million agrarian farmers are involved in growing soya,

http://dx.doi.org/10.3362/9781780448671.012

groundnuts and other ingredients, which are used in poultry feed. Of course, these crops have other uses too. India is the world's fifth largest producer of eggs and ninth largest producer of poultry meat. India is positioned seventh in world poultry production, according to the Ministry of Food Processing Industries in India (MOFPI, 2011).

In 1986, Mr Soundararajan and Mr Sundararajan set up Suguna poultry farm with 200 layer birds at Udumalpet in South India. In 1989–90, when chicken prices crashed because of an over-supply of birds in the local market, Suguna saw an opportunity for business growth by helping other poultry farmers who had bought feed and medicines on credit and could not clear their debts. To help them recover their money, Suguna began to provide feed and veterinary support to indebted farmers in return for the end product: eggs. This model was successful and over a period of time, Suguna evolved a business model around this strategy.

Today Suguna Poultry ranks among the top ten poultry companies worldwide. It is the largest poultry business in India with an annual turnover of nearly half a billion dollars. With operations in 11 states across India, it offers a range of poultry products and services. The fully integrated operations cover broiler and layer farming, hatcheries, feed mills, processing plants, vaccines and exports. Suguna markets live broiler chicken, eggs and frozen chicken. Recently, Suguna has also set up a chain of modern dedicated retail outlets, which sell only poultry items.

Suguna's poultry integration model provides a win–win situation for both the farmer and the integrator. Suguna provides project guidance and assistance in getting finance, as well as training and assistance on poultry management. In some areas, new entrants into the business approach Suguna with only land, and Suguna advises them on how to put up a poultry farm, and then provides chicks and trains the farmers to manage the birds. Farmers are provided with day-old chicks, feed and veterinary support. Performance is monitored, sometimes on a daily basis. The Suguna field staff visit the farms to check on the health of the birds, feed intake, growth and mortality levels. After six weeks, the birds are weighed and are ready to be sold to Suguna and then sold on by Suguna.

Suguna has successfully reduced the number of direct links in the poultry value chain from as high as 14 to 4 or 5. The farmers deal only with the company, and get assured returns. Suguna provides them with all the inputs, and in return, the farmers get a fixed growing charge. Suguna bears the market risks and to a large extent the production risks as well, thereby protecting the livelihoods of farmers. Regardless of the market price, the farmers get the assured growing charge; sometimes, extra incentives are given based on market conditions, in order to reduce the farmers' temptation to 'side-sell' the finished broilers.

Suguna prefers to work with smaller farms than with large farms with 100,000 or more birds because the mortality rate for large farms is much higher.

Also, a disease outbreak can sometimes wipe out an entire batch of birds in one location. With their current model, Suguna can manage these issues. Over time, Suguna has also discovered that the cost of managing several small farms is quite similar to the cost of managing one large farm. They realize also that managing very small farms of less than 2,500 birds is a challenge.

Suguna procures from over 20,000 farmers in 8,000 villages across 11 statesin its value chain. They are supported by over 4,800 full-time Suguna employees. The company claims to have benefited over 500,000 people by direct or indirect employment. To support the value chain, Suguna has adopted a decentralized model, setting up 36 hatcheries producing 395 million chicks per annum and 50 feed mills producing 1.56 million tonnes of feed per annum. That has resulted in a value chain that produces over 10,000 tons of processed chicken meat every year. Even though Suguna started with layers producing eggs, they later shifted to broilers, which are grown for their meat. Only recently, they have restarted layer operations, modelled on their broiler operations, but this has only been rolled out to a very limited extent.

As a result of their partnership with Suguna, a typical farmer's increase in annual income is likely to be between Rs40,000 to 200,000, or $800–4,000. This makes up a substantial proportion of the family's annual income. The labour for their mini-poultry farms is usually provided by the farmer and his family, but many farmers also employ a helper on a monthly salary.

'Poultry Integration', as introduced and pioneered by Suguna in the country, has substantially improved the livelihoods of many farmers in rural India. Several other poultry companies such as Venky's, Godrej Agrovet and others have learned from Suguna's success and have tried to replicate elements in their own chains.

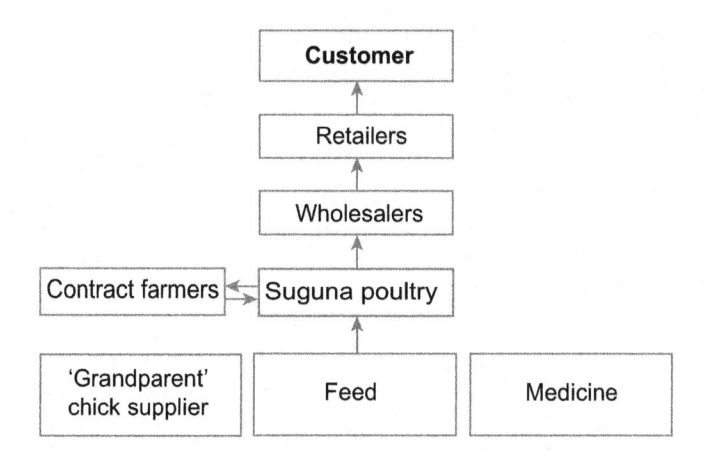

Figure 12.1 Business model of Suguna Poultry

Table 12.1 Value addition in the Suguna Poultry value chain

Value addition	Cost/kilo
Direct cost of inputs (chicks, feed, medicines)	Rs45
Payment to contract farmer	Rs4
Margin for Suguna	Rs11
Sale price to trader	Rs60
Sale price to retailer	Rs70
Price to consumer	Rs90

Note: Prices of poultry and inputs vary over the year, so approximate average prices per kilo are given. US$1 = Rs50.

The major advantages of Suguna's approach to its farmers are: providing better inputs in terms of day-old chicks, feed, and medicine consistently over time; providing on-farm veterinary assistance; and reducing the marketing risks of farmers. The main tasks that are carried out by Suguna are: providing day-old chicks to the farmers; providing feed and medicine along with on-farm assistance; and assured collection of birds and prompt payment of the growing charge.

The basic methodology is strictly for profit, but is not very different from that used by PRADAN, a well-known development NGO, in its efforts to improve the livelihoods of very poor tribal households in and around Kesla, south of Bhopal, in western Madhya Pradesh, albeit on a much smaller scale.

PRADAN's approach to poultry for the poor

PRADAN, one of India's best-known NGOs working in livelihoods, started its poultry operations in Kesla in 1992. Their original plan was to help village women to upgrade and market the existing local birds, which most households kept around their homesteads and sold to raise money for unexpected needs. This was not generally successful, and after several experiments PRADAN settled on caged units of between 300 and 400 birds, which are managed on a similar basis to that employed by Suguna. PRADAN has promoted a poultry cooperative, which is owned by the women. It is known as KPS (Kesla Poultry Samity), and is the largest poultry producer in its area of Madhya Pradesh.

Each family invests about Rs36,000 in a shed, which is financed with a low-cost bank loan under a government-subsidized scheme for tribal people, and the cooperative provides day-old chicks, feed, medicine and litter to its members, who raise between seven and eight cycles a year. The cooperative employs trained field workers to supervise and support its members; one such worker serves between 25 and 30 producers. The cooperative sells to local markets in the area, and also operates its own dedicated retail shops.

By 2009 some 5,300 families were raising poultry in this way; on average they were earning a margin of about Rs5 per bird. This amounts to between Rs9,000 and Rs16,000 a year from their units, or about Rs50 for each day when a batch was in process. This was a very substantial addition to the household income.

We have not been able to obtain any figures for the cost of PRADAN's interventions over the 19-year period, which have been funded with grants from the Ford Foundation and other sources.

The survey

To understand the impact of Suguna's business on the smallholders associated with the company, we undertook a survey of 180 smallholders spread over West Bengal, Odisha, Andhra Pradesh and Tamil Nadu, 4 of the 11 states where Suguna operates. The data were collected from the head of the household and on several occasions also from the spouse of the head of the household.

The results of the survey have been collated and are presented in the following sections.

Changes among smallholders

Poultry farming is a major livelihood for all the respondents. The households involved in poultry farming are very diverse as most of them were involved in various other livelihoods as well.

The farms differed quite widely with respect to the size of their poultry units; the mean number of birds in the sample was 4,400, with a standard deviation of 2.955. The majority of the farmers who were interviewed own a poultry farm with less than 6,000 birds, and most owned farms of between 2,000 and 4,000 birds. Less than 10 per cent of the sample had more than 8,000 birds.

In our sample, the average time that the farmers had been associated with Suguna was just under four years. Most of the farmers had been in the chain for between three and five years. This reflects how Suguna has expanded in the last five years. However, we should remember that we deliberately avoided interviewing farmers who had been in the value chain for less than three years.

Of all the four states in which the data were collected, the average farm size was highest for Tamil Nadu with 6,500 birds and the lowest was West Bengal with 2,430 birds. The farm size is affected mainly by the capacity of the farmer to invest in a new business. In Andhra Pradesh and Tamil Nadu, Suguna encourages the farmers to have more than 3,000 birds. It is also compulsory for the farmers to have their own basic infrastructure such as motorized water pumps, water, feeders and drinkers.

Every form of enterprise needs an investment to start, and this is a major barrier for many people. The same is the case with poultry. Typically the initial infrastructure cost is about Rs75 to Rs90 per square foot; a farm of 2,500 square feet needs an initial investment of about Rs200,000.

The maximum number of farms in our sample were between 2,000 and 4,000 birds, which involved an investment of between Rs200,000 and Rs400,000. Only a very small number had invested less than Rs100,000, either because they had very small units or because they were able to use existing facilities.

The major factors that influence the farmers to take up poultry, in addition to other farming activities, are risk diversification and the shorter cash cycle of two months. In this context most of the poultry farmers prefer to operate their operation on a fairly small scale, with between 2,000 and 6,000 birds. These small units allow the farmers to reduce operation costs by using the minimum of external paid labour. This is done either by working themselves on their farm or by using the same labourers for their crops and their poultry.

Over a period of time some farms have expanded, but the number is not significant. In most cases farmers prefer not to expand their poultry operation, but rather to expand in other types of agriculture. The farmers believe that larger poultry farms require much more hard work, without commensurate rewards.

The major sources of funds are loans, family inheritance and savings. The loans were both from informal sources and from banks. Although many farmers mentioned only a single source of finance, it was evident that most had used multiple sources.

Suguna purchases the grown broilers and pays the growing charges to the farmers. It is mandatory that the chicks are sold only to Suguna as mentioned in the contract. The growing charge given by Suguna varies depending on the total body weight of birds, and the mortality and feed conversion rates.

The final cost is derived based on the actual cost of production for one kilo of meat. The difference between the farmers' cost of production and the baseline production value is noted. The growing charges are based on the difference between these two figures.

The majority of the farmers who were surveyed were earning between Rs50,000 and Rs200,000 per annum from their Suguna poultry operations. The average income was higher for larger farms, but the proportionate return on investment remains around the same. There was a fairly close correlation between farm sizes, farmers' investments and their earnings. This shows that Suguna has successfully standardized the process so that it can offer its farmers a reasonable return for their time and money, but can also ensure that its supplies of poultry cost as little as possible.

For small-scale farmers, the initial investment in a poultry unit is a heavy commitment, particularly if it is financed with a bank loan. It is therefore important that their units break even as soon as possible. Most of the farmers stated that they were able to achieve break-even in less than four years, and some were breaking even in less than two years. Since there is a significant number of farmers who have spent between two and four years in the Suguna

Table 12.2 Economics of a 2,500-chick poultry farm supplying to Suguna

Item	Amount
Investment	Rs230,000
Number of chicks per batch	2,500
Survival rate (%)	98
Average weight per grown chicken (kg)	1.7
Total weight at culling (kg)	4,165
Average growing charge (Rs/kg)	Rs3.80
Revenue from birds	Rs15,827
Less: electricity	Rs1,000
Less: consumables (rice-husk/sawdust)	Rs1,500
Net revenue per batch	Rs13,327
Number of batches per year	6
Total revenue per year	Rs79,962
Payback period for original investment	2.88

Note: US$1 = Rs50

value chain, it can be inferred that a large number of the businesses have been able to meet break-even in their short association with Suguna.

The range and diversity of rural households' livelihood portfolios is an indicator of the risk-taking capacity of the household, especially for families below the poverty level. Diversifying the livelihood portfolio helps the poor to decrease the vulnerability of the household to external risks such as drought, floods, prices, and other economic changes.

In 2010 there was a strike by farmers in Tamil Nadu that resulted in an increase in the growing charges. Some farmers said that they would like to form a type of Suguna producers union to enable them to express their opinions. Similar kinds of activities were not found in other regions, and in general the high satisfaction level means that there is little reason for the farmers to come together in order to present a common front and demand improved conditions.

The analysis of the livelihood portfolios of the farmers indicates that the prime reason for their association with Suguna was to increase their income by livelihood diversification. Their poultry business forms a part of the livelihood portfolio, but at the same time a major significant portion of their physical and natural assets such as land is being used for farming. As their association with Suguna becomes more profitable, many people have also reduced or stopped their earlier less profitable ventures or have been able to diversify their livelihood portfolios by starting new ventures. There are some cases of reverse migration where the farmers returned to their native villages leaving waged jobs in cities in order to start raising poultry for Suguna. This has a positive impact on society in general by bringing back relatives to their families.

In our survey 16 per cent of the poultry farmers had already been associated with the poultry industry before they joined Suguna, of whom about 20 per cent were in the poultry business independently, with no tie to any other company.

The Progress out of Poverty Index (PPI) was used to measure farmers' poverty levels, and the change that had happened during the time they had worked with Suguna. Table 12.3 shows the relationship between the PPI scores of the households before joining Suguna and the current levels. There is a clear shift in the number of households from a lower score to higher. This does not of course prove any causal relationship between working with Suguna and changes in their levels of poverty, because the farmers may have been the type of people who would have improved their position in some other way, or their position may have improved as a result of incomes from other sources, or there may have been a general improvement in the economy from which all farmers benefited. The average PPI score of the households has increased to a level of 59.2 from 49.7 (see Table 12.3).

Of the 181 farmers interviewed, only 16 had employees looking after the poultry farm. Ten of the employees were interviewed. The main advantages of working in a poultry farm were the fact that it provided continuous employment through the year and it was physically less taxing than working as an agricultural labourer. All the employees were men; their ages varied from 18 to 62.

Table 12.3 Farmers' PPI score before joining Suguna Poultry and at current levels

PPI Score	Number of farmers with score before joining Suguna value chain	Number of farmers with score after joining Suguna value chain
Under 10	1	0
10–20	8	4
21–30	15	6
31–40	40	17
41–50	47	35
51–60	24	43
61–70	10	24
71–80	18	25
81–90	9	14
91 and over	8	12

Table 12.4 Characteristics of poultry farm employees

Employee characteristics	Amount
Average salary	Rs1,300
Average number of working hours per week	68
Average score on PPI	32
Average number of family members	5.5
Average years working in poultry farm	2.1

Box 12.1 Poultry farmer case study

Mrs Rajaveni has a poultry farm of 3,500 birds in Thungavi village, Tamil Nadu. She and her husband manage the poultry farm. They have two children, a daughter of seven and a baby son of ten months. They do not need any full-time wage labourers for their poultry operation as it is small enough for them to do the work themselves. Apart from poultry they also participate in regular agriculture. They have 1 acre of land on which they produce beans and other vegetables. They also have two cows that augment their family income.

Previously her husband was working as a computer operator in a parcel service company in Kerala. They decided to start a poultry unit in their village since they had heard that it was very profitable. Her husband quit his job and came back to their village, and revenue from poultry is now the family's major source of income. On average they turn over 6 batches in a year, and they grow each batch of chicks for about 36 days. Suguna provides them with day-old chicks, feed, medicine and vaccines. Their major costs are for coal and electricity to keep the chicks warm, and they have two pumps to provide a regular supply of water.

They invested Rs400,000 in the shed for their poultry farm, and a further Rs7,000 for drinkers and feeders. They started the business in 2006, working with Shanti and VKS, two other poultry firms. Rajaveni switched to Suguna in 2008, and she rates Suguna as a better company than the other two. The major points of difference are their technical services, and timely supply of feed, chicks and medicines.

They are now earning an average net income of about Rs20,000 per batch, or some Rs120,000 a year, and Rajaveni says that this has substantially increased her family's income. At the same time she says their consumption needs have also increased. Her household's score on the PPI index has marginally increased during the period, from 63 to 70. Overall she is highly satisfied with Suguna and she hopes that the company will keep up its good work.

References

MOFPI (Ministry of Food Processing Industries) (2011) *Poultry and Products Annual Report*, Delhi: MOFPI.

MOFPI (2013) *Annual Report 2012–13*, Delhi: MOFPI.

MSPI (Ministry of Statistics and Programme Implementation) (2011) *National Sample Survey* (67th round), Delhi: MSPI.

Seetharaman, G. (2013) 'Rs40,000 crore poultry meat market: Indians prefer chicken; mutton, beef far behind', *The Economic Times*, 27 October.

About the authors

Malcolm Harper is Emeritus Professor of Cranfield University and has attempted since 1970 to make basic management and business tools work for the alleviation of poverty.

Rajeev Roy was an entrepreneur and currently teaches entrepreneurship. He has been engaged in several entrepreneurship development initiatives across the world. He is involved in mentoring several start-ups in India.

Phanish Kumar is a supply chain professional. He has an MBA and a degree in Biotechnology. He has worked with ITC and Infosys.

PART THREE
Non-commodity foods

CHAPTER 13
Green beans – from small farmers in Senegal to gourmets in Europe

Miet Maertens

Abstract

Exports of fresh fruits and vegetables (FFV) from developing countries are increasing rapidly and can raise rural incomes because of the link with the rural economy, the high value of the produce and the labour-intensive production systems. Many poor countries therefore pursue the development of FFV export chains as a poverty-reduction strategy. Donors, local private companies, foreign investors and large multinational companies increasingly invest in FFV export businesses in developing regions. This case examines bean exports in Senegal. In order to guarantee more stringent demands from importers on product quality, safety and traceability, bean exporters tend to replace contract farming with their own, large-scale estate production. This has increased the demand for workers on the estate farms and in the processing centres of the export companies. The author concludes that particularly poorer families have benefited since they are more likely to get these jobs. Therefore, the main poverty-reducing effects of bean exports from Senegal have been realized through the labour market, specifically through the creation of jobs that are accessible to the poorest rural households, rather than through the inclusion of smallholder producers in the export chain through contract farming. This challenges the implicit assumption that export supply chains need to integrate smallholder suppliers if export expansion is to benefit the rural poor. The growth of high-value export production, even if it is from large-scale estate farms, should be seen as a source of employment for the rural poor and not as excluding smallholder producers.

Keywords: green beans; Senegal; exports; employment; contract farming

Background

Exports of FFV (fresh fruit and vegetables) from Senegal increased from US$0.78 billion in 1995 to US$44.64 billion in 2011 (see Figure 13.1). The main export crops are tomatoes (28 per cent of total exports), green beans (21 per cent), melons (24 per cent) and, more recently, mangos (11 per cent). Senegal has developed into one of the main fresh produce exporters in Sub-Saharan Africa. The main market is the European Union, particularly France, Netherlands, Germany, United Kingdom, Spain and Belgium.

http://dx.doi.org/10.3362/9781780448671.013

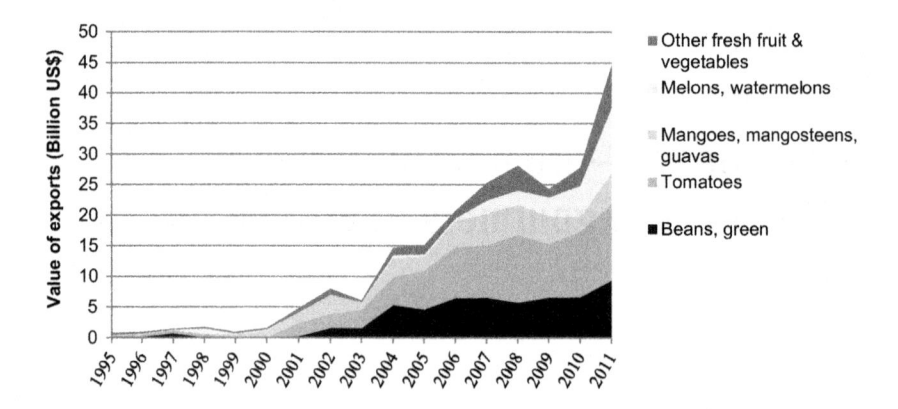

Figure 13.1 Value of fresh fruit and vegetable exports from Senegal, 1995–2011 (US$ billion)
Source: Based on FAO statistics

Senegal is now the fourth largest African supplier of mangos to the EU, after Cote d'Ivoire, Mali and Burkina Faso, and the fourth largest supplier of green beans after Morocco, Kenya and Egypt.

The boom in FFV exports fits Senegal's strategy of agricultural export diversification towards higher-value commodities. This strategy was adopted after the devaluation of the franc CFA (Communauté Financière Africaine) in 1994, and following decades of dependence on groundnuts as the main agricultural export commodity. Export revenues collapsed as a result of falling world groundnut prices in the 1980s, leading to a major crisis in the sector and a reorientation towards FFV exports. FFV now constitutes 10 per cent of total agri-food exports in Senegal, while groundnuts represent only 2.3 per cent. The government in Senegal has played an active role in attracting foreign investors in FFV exports. Since 1995, FFV companies that export at least 80 per cent of their output receive the Free Export Enterprise statute, which provides tax reductions and exemptions. The government also invested in infrastructure and institutions that benefit FFV exports, for example cold storage facilities at the airport and the main harbour in Dakar, and laboratory testing of food quality and safety. More recently, the FFV export sector also received some donor support. Since 2006, some smaller FFV export companies in Senegal received assistance from the ColeACP-PIP (Comité de Liaison Europe-Afrique-Caraïbes-Pacifique Programme de coopération Européen pour un développement durable du secteur fruits et légumes de ACP) programme, financed by the European Union. In 2007 the Program for the Development Agricultural Markets in Senegal (PDMAS) started with several donors, mainly the World Bank.

This case is about bean exports. Green beans are produced almost exclusively for export. Domestic trade is limited to very small volumes, mainly produce that does not comply with export standards and requirements, and goes to urban markets, particularly restaurants and hotels. Beans are mainly sourced from the Niayes region stretching along the coast north of Dakar (see Figure 13.2). Due to its specific climatic conditions, this region is well suited to horticultural production. But it is a densely populated region where agricultural land is becoming scarce and, because of salinity, access to water is an important constraint for agriculture.

The bean export season is from December until April. From an agronomic point of view, vegetable production would be optimal during the July to October rainy season but during these months demand for bean imports in the EU is low because of local production. Thus, Senegal exports to the EU when there is limited EU production and when its competitors, Morocco, Kenya and Egypt

⊚   City

    Selected rural communities

Figure 13.2 Bean export region in Senegal and studied communities
Source: Atlas du Sénégal – IRD – Cartographie A. Le FUR-AFDEC

face less favourable climatic conditions. Next to beans for export during the export season, most households in the Niayes region are smallholder farmers producing a large variety of vegetables and staples for the local market and for their own consumption.

During the past two decades, food quality and safety requirements have increased rapidly. Regular controls on the safety, labelling and marketing standards of imported vegetables are carried out at the point of entry to ensure they comply with all EU requirements. Certain exporting countries are allowed to perform the conformity checking operations themselves, prior to entry in the EU. Since 2006, Senegal is among the few African countries, along with Kenya, South Africa and Morocco, that is accredited by the EU to do this. It reduces exporters' risks of non-conformity and entry denial of produce.

In addition to increasing public standards, many large companies have established their own food standards that are even stricter, related to food quality and safety, or environmental and ethical concerns. The most important private standard for vegetable exports from Senegal to the EU is GlobalGAP. This consists of a series of protocols for traceability and the application of good agricultural practices. Agri-food businesses in the EU increasingly require such private certification from their suppliers. By 2010, more than 40 retailers including the largest in 15 countries, mainly in Western Europe, required their suppliers to be GlobalGAP certified. As a result, GlobalGAP is spreading in developing countries and to bean exports in Senegal.

Despite these increasing standards, Senegal has been able to increase FFV export earnings, which proves that tightening legislation does not necessarily undermine the competitive position of developing countries in international agricultural markets. The development of a certification scheme and validation of the label 'Origine Sénégal' played an essential role in raising the standards of Senegalese fruits and vegetables, and increasing exports.

The structure of the bean export chain

The following description of the bean export value chain is based on interviews and data from many sources. These included local experts from two exporter organizations SEPAS (Syndicat des Exportateurs des Produits Agricoles) and ONAPES (Organisation National des Producteurs Exportateurs de Fruits et Légumes de Sénégal), from the Directorate of Horticulture, the Centre for the Development of Horticulture, and others. Information was also provided by 9 bean export companies in the Niayes region in 2005 and by 12 companies in 2010. These companies together represented more than two-thirds of bean exports in both years.

The bean export value chain is depicted in Figure 13.3. This figure represents the situation in the bean export sector in Senegal in 2005, when the majority of the data were collected.

Supply chain diagram

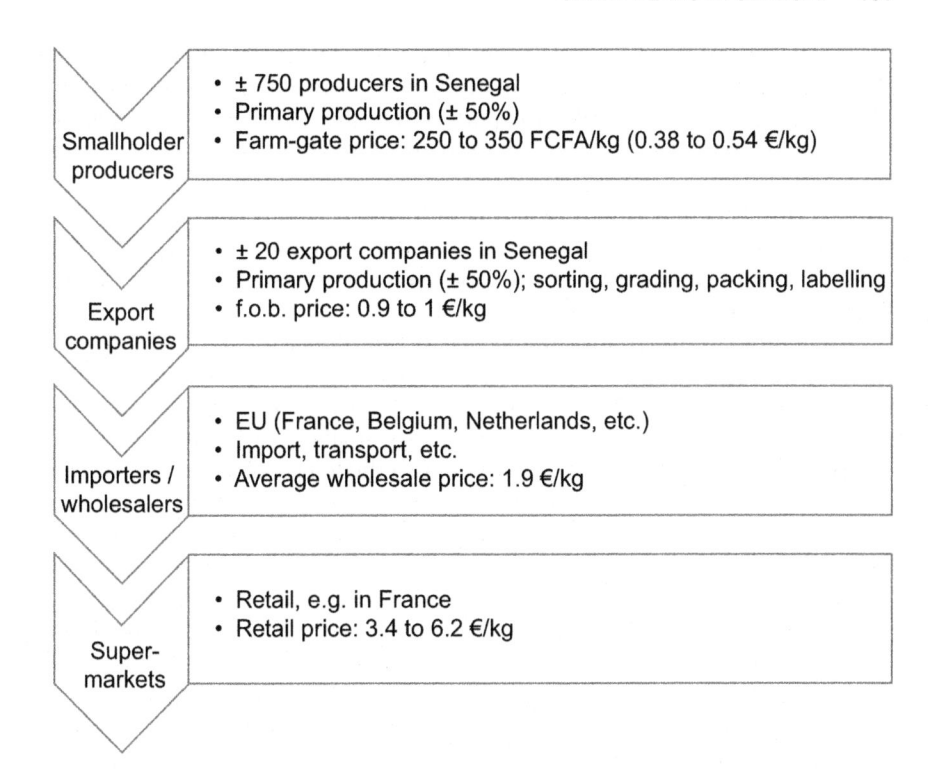

Smallholder producers
- ± 750 producers in Senegal
- Primary production (± 50%)
- Farm-gate price: 250 to 350 FCFA/kg (0.38 to 0.54 €/kg)

Export companies
- ± 20 export companies in Senegal
- Primary production (± 50%); sorting, grading, packing, labelling
- f.o.b. price: 0.9 to 1 €/kg

Importers / wholesalers
- EU (France, Belgium, Netherlands, etc.)
- Import, transport, etc.
- Average wholesale price: 1.9 €/kg

Supermarkets
- Retail, e.g. in France
- Retail price: 3.4 to 6.2 €/kg

Figure 13.3 Diagram of the bean export supply chain from Senegal (situation in 2005)
Source: Based on data from interviews and survey, and from INSEE
Note: One equals approximately US$1.25.

Export companies and consolidation

When bean exports started to increase sharply in 2000, there were around 25 export companies exporting fresh beans to the EU. They included larger and smaller companies, with local and foreign investment. However, with the introduction of the new EU Food Law of 2002 and the spread of GlobalGAP, the bean export sector has been consolidating. The number of exporting companies fell from 27 in 2002, to 20 in 2005 and to 14 in 2010, with mainly smaller exporters dropping out. Moreover, the market share of the three largest companies increased from less than half in 2002 to two thirds in 2005. This means that there is a sharp consolidation at the level of exporting in the chain.

Exporters are forced to stay up to date with the changing legislation and make additional investments in order to comply with food standards. Many exporters face financial constraints when they invest in upgrading their capacity to conform with EU regulations and private standards. Only large, and particularly partially foreign-owned companies with access to foreign capital, are able to make the necessary investments. Many smaller exporters have left the market in recent years because they cannot afford the cost. In 2005, the first bean exporter in the Niayes region obtained GlobalGAP certification and another six large exporters were aiming to become GlobalGAP certified. By 2010, five bean export companies had obtained the certification. The companies see GlobalGAP as effectively essential to enter the EU market. These firms have made substantial investments to comply with the private requirements, including cold storage facilities and transport, facilities for selection and packaging, control mechanisms, improvements of sanitary conditions at the conditioning centre and others.

There are some important differences between GlobalGAP certified and non-certified companies. Certified companies have larger yearly export volumes of beans, averaging 900 to 1,000 tonnes, and with a larger market share, on average 17 per cent. Non-certified companies have smaller yearly export volumes of 200 tonnes and smaller market shares. Also the length of the export season differs. Certified companies export beans to the EU on average five months a year while non-certified companies export on average only three months a year. These figures may suggest that certification may improve a company's international competitiveness and reputation – or that the best performing companies become certified. Certification may lead to new markets, new buyers and more stable contracts with buyers. As certification reduces uncertainty and trade costs, this can result in increased and more stable export volumes and a longer export season. Not surprisingly, certified companies have on average larger shares of foreign capital (50 per cent) than non-certified companies (12 per cent), which indicates the importance of foreign capital to make necessary investments for compliance with standards.

Vertical coordination and ownership integration

The increasingly stringent EU public and private standards have had major consequences for the organization and the structure of the bean export sector in Senegal. Higher food standards increase the need for tighter coordination and lead to important changes in the governance system of food export supply chains.

The spread of GlobalGAP has contributed to an important restructuring of the supply chain. This is apparent in the increasing degree of vertical coordination at different levels of the chains. Vertical coordination increased, both downstream and upstream. Downstream, this includes tighter coordination between exporters and overseas buyers. In recent years, many bean exporters, especially larger certified exporters, have shifted from loose agreements with downstream EU importers and wholesalers to more binding contracts. Smaller exporters deal with importers through non-binding indicative agreements on the supplied quantity. Exporters say that the volatility of EU market prices and produce refusals by importers mean that tighter coordination is necessary.

In addition, in order to guarantee product quality, safety and traceability throughout the supply chain and to ensure accurate timing of production and harvesting, bean exporters work more tightly with suppliers. This occurs in two ways. The first is through more elaborate production contracts with local suppliers and tighter coordination within those contracts. Contracts signed with smallholder farms are typically specified for one season, lasting from November to April, and indicate the area to be planted, usually 0.5-1 hectare, all the technical requirements and the price. As part of the contract, the firms provide technical assistance and inputs to the farmers, especially seeds and chemicals, sometimes also cash credit. Some firms take over the complete management of fertilizer and pesticide application and daily or weekly inspection of the farmers' fields by company agents and technicians. Field preparation, planting and/or harvesting can be coordinated and financed completely by the contractor firm. Larger exporters in particular provide pre-financing and control farmers more tightly, while smaller exporters leave management decisions to the farmers.

A second, and even more radical, change toward vertical coordination is the shift from smallholder contract-based farming to large-scale estate production. Larger exporters, and especially exporters seeking to become GlobalGAP certified, are increasingly producing on fully integrated estates. In 2000, after the introduction of EurepGAP, the predecessor of GlobalGAP, the seven largest FFV export companies in Senegal founded a new exporters, organization, ONAPES, in addition to the already existing exporters' organization SEPAS. One of ONAPES' members' aims was to comply with traceability standards and become EurepGAP certified. As part of this, the ONAPES exporting companies agreed among themselves that each member should seek to export 200 tonnes every season, of which at least half should originate from the company's own estate production. This measure had a profound impact on the structure of the export chain. Certain firms decreased their sourcing from local suppliers from 100 per cent of total export volume to 60 per cent or even 20 per cent in a couple of years. These companies cited quality rather than quantity to be the reason for this change. This shift from contract farming with local suppliers to vertically integrated production has resulted in a sharp decrease in the share of export produce that is sourced from smallholder suppliers. While in 1999 almost 100 per cent of beans were sourced from smallholders under contract, this fell to around 50 per cent by 2005 and slightly dropped further to around 45 per cent in 2010. Around 750 smallholder farmers are supplying beans to export companies under contract-farming arrangements with the companies. While their importance in overall bean export production has decreased from 100 per cent to less than 50 per cent, the number of contracted smallholders in the sector has decreased less dramatically as total exports are continuously increasing (see Figure 13.1).

Agro-industrial production and employment

The shift in sourcing from contracted local suppliers to vertically integrated production has resulted in agro-industrial production by export companies themselves. Export companies that shift towards vertical integration and their own primary production do this with leased land, often leased for 99 years. Land is allocated to them by the community councils who hold the authority over land. These councils negotiate directly with the companies over the access to land, and lease contracts often include, as well as the price, the provision of social benefits by the company, such as investment in roads, health or schools, and preferential access to employment in the company for households from the community. The land lease deals usually cover land that is not intensively used by local farmers; the companies often have to invest heavily in irrigation and other infrastructure to make the land productive. A survey of households in the Niayes region (see below) found no farmers who had been displaced from their land as a result of FFV export companies expanding their own production. As land and water become increasingly scarce in the Niayes region, FFV export companies seeking to expand their businesses increasingly do so in other regions of Senegal, such as in the Senegal River Delta region, where both land and water are more abundantly available.

The shift towards agro-industrial production has resulted in increased demand for labour on the export companies' fields. In addition, the need for labour-intensive post-harvesting and processing increased because of increased requirements for sorting, grading, washing and labelling in public regulations and private standards. By 2010 more than 12,000 workers were employed by bean export companies. These workers mainly come from farm households that allocate some of their family labour to waged employment in the bean export sector, mainly during the agricultural off-season, and some to the household's own farm and other business activities. As mentioned above, none of these workers were displaced from their land. Rather, wage work in the bean export industry became a complementary off-season income-earning activity for farm households.

Rural households in the bean export supply chain

These changes in the structure and organization of vegetable export chains, especially the shift from smallholder contract farming to large-scale agro-industrial production, have major implications for rural households in the producing areas. A household survey was carried out in three rural communities in the Niayes region. The majority of households in this area are smallholder farmers producing beans for export and also a large variety of vegetables and basic food crops for the local market and for direct consumption. The first survey was carried out in 2005 and covered 300 farm households in 25 villages in the communities. These 300 households were re-interviewed in 2010, along with another 150 households in 15 villages who were added to the sample. In between 2005 and 2010, only five households dropped out of the survey. The original sample of 300 households in 2005 includes 59 contract farmers who supply beans to the export companies, and the extended sample of 450 households in 2010 includes 73 bean contract farmers. These households have been oversampled specifically to analyse the involvement in the bean export supply chain. The sample is representative of small household farms in the area. The average farm size is 5 hectares and 88 per cent of the sampled households cultivate less than 10 hectares – which in the region is considered the smallholder threshold. The sampled households have different farm and off-farm income sources, but farming constitutes on average more than 80 per cent of total household income.

As exports of green beans increased in Senegal in the past decade, the participation of local rural farm households in the export supply chain also increased. In the sampled villages, the total participation of local households in bean export production increased from less than 15 per cent in 1996 to 40 per cent in 2005 (see Figure 13.4). However, the way in which these households are involved in the export chain has changed drastically over the years. As a result of changes in the sourcing strategies of large exporting companies, the share of local households supplying beans to the export industry in contract-farming arrangements dropped sharply after 2000. From 1996 to 2000, the proportion of contract farmers in the surveyed villages increased from 10 per cent to 22 per cent. However, from 2000 to 2005, this dropped to less than 10 per cent and to less than 5 per cent by 2010. In the sample, 72 per cent of contracted bean farmers lost their contracts with the export industry in the period 2000–05, falling back on production for local markets and other income-generating activities. However, the share of local households with jobs in the fields and processing plants of agro-exporting companies increased from 6 per cent in 2000 to almost 35 per cent in 2005, and to 40 per cent by

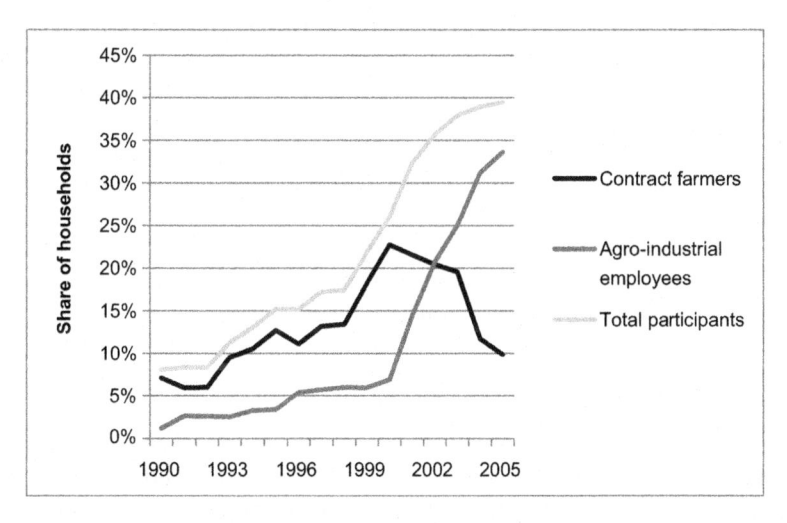

Figure 13.4 Participation of local households as contracted suppliers and agro-industrial employees in the bean export chain, 1996–2005
Source: Maertens and Swinnen, 2009

2010. Almost half of the previously contracted farmers who lost their production contracts got these jobs. Also other farm households, who were previously not involved in bean exports, started to get this work from 2000 onwards. The large expansion of employment in the export sector from 2000 onwards is related to the shift from smallholder contract farming to agro-industrial estate farming, and to the expansion of labour-intensive post-harvest handling and processing of produce before exporting.

Around 750 smallholder farmers and 12,000 workers participate in the bean export chain; the former as suppliers of produce in contract farming schemes and the latter as workers on the estate farms and in the processing centres of the export companies. Most employees are recruited on a day-to-day basis during the export season. The jobs consist of work in the fields, such as planting, fertilizer and pesticide application, harvesting, and work in the processing plants, such as sorting, packaging and labelling produce. It must be noted that households in this area generally only allocate part of their land and labour to these activities, and that they primarily remain independent smallholder farm households producing for the local market.

The poverty-reducing effects of expanding exports critically depend on the type of households included, either as contracted suppliers or as agro-industrial employees in the export chains. In Table 13.1 the characteristics of three groups of households are compared: those working as employees in the green bean export industry; households supplying green beans to the export industry as contracted farmers; and households who do not participate at all in green bean export production. From this comparison, it is clear that those households whose members work as employees are larger households, with more workers and less land, livestock and fewer non-land assets than those who do not participate in the bean export sector at all. They are relatively poorer than the other households. Households who are contracted to grow beans are relatively larger and better off in terms of land, livestock and non-land assets than households that do not participate. Contract farmers also have a slightly higher level of education, and employees and contract farmers also have a slightly higher level of social capital, as measured by their membership of community organizations.

These results indicate that relatively larger farms and better-off households are more likely to participate in contract farming, while poorer households are more likely to get jobs with agro-industrial firms.

Income and poverty effects

The development of the bean export sector in Senegal has had an important impact on the incomes of local farm households and on poverty reduction in the Niayes region. Contract farming to supply beans to export companies and employment in agro-industrial export companies has resulted in significantly higher incomes. Based on the survey data, it is estimated that contracting with the export sector leads to incomes that are 110 per cent higher than the average

Table 13.1 Comparison of household characteristics

	Non-participants	Employees in the bean export chain	Contract farmers in the bean export chain
Number of households in the sample	158	109	59
Age of the household head	53	56	53
Number of labourers[1]	6.4	7.7	7.7
Dependency ratio[2]	0.571	0.566	0.527
Female headed households (%)	3.3	2.8	0
Household head with primary education (%)	16.5	18.8	19.4
Ethnicity (non-Wolof[3]) (%)	31	17	32
Membership of an organization (%)	54	62	77
Per adult-equivalent landholdings (ha)[4]	0.84	0.78	1.03
Livestock units[5]	2.87	1.87	4.14
Non-land assets (1000 FCFA[6])	320.9	176.9	308.8

Note: [1]Labour is measured as the number of persons who are older than 15 and able to work. [2]Dependency ratio is calculated as the number of dependents (children below the age of 15, students and those unable to work) over the total household size. [3]Non-Wolof households refer to ethnic minorities in Senegal. [4]Per adult-equivalent measures are calculated using the modified OECD (Organization for Economic Co-operation and Development) adult equivalence scales. A weight of 1 is assigned to the first adult, 0.5 for all other adults and 0.3 for children. [5]One livestock unit equals 1 cow/horse, 0.8 donkey and 0.2 sheep/goat/pig. [6]US$1 equals around 510 FCFA.
Source: Maertens and Swinnen, 2009

income in the region while for jobs in the export industry this is 60 per cent. The observed shift in supply chain structure from smallholder contract farming to large-scale vertically integrated estate production has resulted in a stronger poverty-alleviating effect. This is the case because the poorest households mainly participate and benefit through employment while participation in contract farming is biased towards relatively better-off households.

Income effects

A comparison of mean incomes across households reveals that average incomes for households involved in vegetable export production are substantially larger than for other households (see Figure 13.5). Farmers who supply beans to export companies on a contract basis have incomes that are more than three times higher than those of households who do not participate in the bean market. In addition, the incomes of contract farmers have increased faster in the period 2005–10 than the incomes of other farm households in the region. The average income of households who have at least one household member working as an employee in the green bean industry is more than two times that of non-participating households, despite the fact that these are households with less land and less assets. In addition, an important share of income, about 30 per cent, for agro-industrial employees comes from these wages. The data suggest that contract farming increases household income by 120 per cent and employment in bean export companies by 60 per cent. Hence, the green bean export supply chain has contributed importantly to raising rural incomes in the Niayes region.

These effects on income are direct and indirect; there are many benefits from contract farming. Farmers can access an export market, in which they can receive higher prices than at home. In 2005, contract farmers usually received between 250 FCFA/kg and 300 FCFA/kg, or sometimes 350 FCFA/kg, for the beans they supply to the export companies (US$1 equals around 510 FCFA). When they sell beans to local traders, they receive 230 FCFA/kg. In addition, contracted famers receive assistance from the contractor companies in the form of inputs, credit, technical extension and managerial advice. This primarily benefits their bean production for export but technical knowledge and modern inputs may also spill over to other crops that farmers produce for the local market. Contract farmers themselves do not see the higher prices and higher incomes as the main benefit from contract farming: 22 per cent of the farmers say that the main benefits from contract farming are their guaranteed

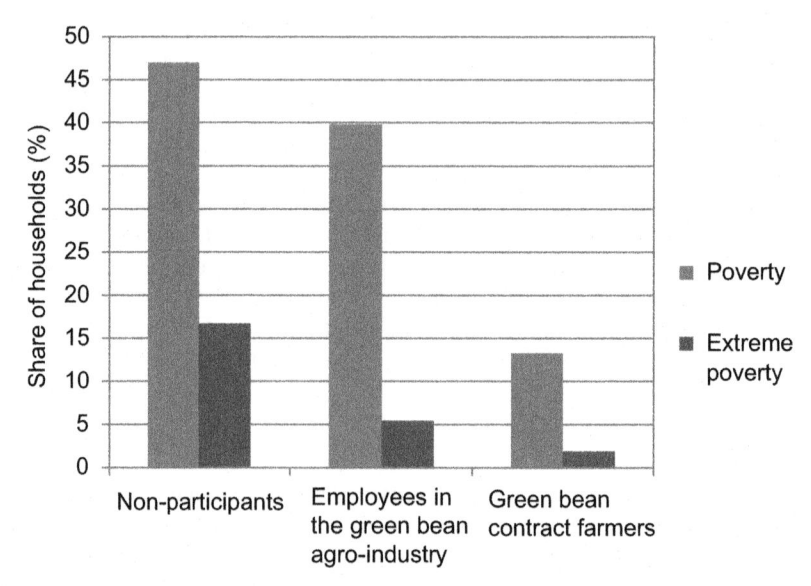

Figure 13.5 Comparison of average farming household income from different sources
Source: Maertens and Swinnen, 2009

outlet to the export market, while 26 per cent mentioned access to credit and inputs.

Household income is calculated as the annual income for the 12-month period prior to the survey in July 2004–05. Farm income takes into account total production valued at market prices in the three different growing seasons, the cost of variable inputs and hired labour, and the depreciation of machinery and equipment. Agricultural wage income includes wages earned in the export agro-industry (80 per cent) and other agricultural wages (20 per cent). Income from non-agricultural sources includes wages from non-farm jobs and income from small businesses, such as small shops and non-labour income such as remittances.

Jobs in export agro-industry create direct as well as indirect benefits. Employees in the bean export industries earn between 1,500 FCFA and 2,000 FCFA or about US$4 per day, depending on the type of activity and the company they work for. The wages in GlobalGAP certified companies are slightly higher than in non-certified companies. For field work, a worker receives on average 1,500 FCFA a day in non-certified companies but 1,800 FCFA in a certified company. The wages that agro-industrial employees in the bean export sector receive add directly to household income. These wages account for about 30 per cent of total household income, on average for those households where at least one household member is employed in the sector. This is a substantial contribution.

The development of bean exports has also been associated with increased job opportunities for women. This has had an impact on female empowerment in the region. About 90 per cent of the employees are female. Males tend to get the few higher-paid permanent jobs but women dominate the far more numerous part-time jobs. They are not paid as well as men, but the gender wage gap in the bean export industry is three to six times lower as compared to other sectors. Hence, bean exports are contributing to a reduction, albeit not an elimination, of direct and indirect gender discrimination in the rural labour markets.

The wages earned in the bean export industry are at least partially invested in the households' farm and off-farm businesses. For example, access to wages from the bean export sector has a positive effect on farm intensification and leads to increased use of modern inputs such as mineral fertilizer and improved seeds, and thereby improves yields and production of local food crops.

In summary, it is not only vegetable contract farmers with more labour, land and capital endowments who have the highest incomes, employees in the vegetable export industries also have incomes that are substantially higher than other households, although they are poorer households with less land and fewer capital endowments.

Poverty reduction

The income effects translate into important poverty reduction effects, as revealed by the proportion of households living on incomes that fall below the national rural poverty lines. The overall incidence of poverty in the research area is estimated to be 42 per cent, which is considerably below the national rural poverty rate of 58 per cent. Both poverty and extreme poverty is substantially lower among households participating in the bean export chain compared to non-participating households (see Figure 13.5). Among contracted green bean farmers, poverty is only 13 per cent and extreme poverty only 1 per cent. Among households employed in the green bean export industry, the incidence of poverty is 40 per cent and of extreme poverty 5 per cent, which is substantially lower than the rate of poverty (76 per cent) and extreme poverty (17 per cent) among non-participating households.

The incidence of poverty and extreme poverty is larger among households that are employees in the bean export industry than among households that are contract farmers in the bean sector. Yet, it is through creating employment opportunities that the bean export sector has had the largest impact on poverty. While contract farming has a large effect on households' income, contract farming is only accessible for farmers who are better endowed with land and non-land assets. The effect on household income from employment in the bean export industry is lower, but these jobs are accessible for poorer households. The income and poverty figures suggest that the vegetable export industry has created possibilities for households with lower land and non-land assets to escape from poverty through the creation of employment opportunities for low skilled labour.

Conclusions and implications

This case shows that the sharp growth in vegetable exports in Senegal has made an important contribution to improving rural incomes and reducing rural poverty in the country. Despite increasing consolidation in global food supply chains, the increasing dominance of multinational holdings in these chains, and the higher standards demanded of fresh produce imports in high-income markets, the expansion of vegetable exports and of high-value horticulture crops in general is an important agricultural and pro-poor development strategy in poor countries. Although the benefits from vegetable export growth in Senegal are concentrated in specific rural communities such as the Niayes region from where the majority of beans are exported, and not shared equally all over the country, other regions and communities of the country will get higher incomes. There is scope for similar pro-poor rural development through the development of high-value and high-quality export supply chains throughout Sub-Saharan Africa.

The main poverty-reducing effects of bean exports from Senegal have been realized through the labour market, specifically through the creation of jobs that are accessible to the poorest rural households, rather than through the inclusion of smallholder producers in the export chain through contract farming. This challenges the implicit assumption that export supply chains need to integrate smallholder suppliers if export expansion is to benefit the rural poor. The growth of high-value export production, even if it is from large-scale estate farms, should be seen as a source of employment for the rural poor and not as the exclusion of smallholder producers.

Reference

Maertens, M. and Swinnen, J.F.M. (2009) 'Trade, standards and poverty: Evidence from Senegal', *World Development* 37(1): 161–178.

About the author

Miet Maertens is Associate Professor in Agricultural Economics at the University of Leuven, Belgium. Her research focuses on sustainable food supply chains in developing countries.

CHAPTER 14

Odisha cashew nuts to global markets – value added all the way

Kulranjan Kujur

Abstract

The cashew industry in India supports the livelihoods of a large number of people and a significant proportion of them are poor. This case examines a value chain that includes both small-scale producers, who have benefited from the global rise in cashew prices, and labourers in processing units that convert raw cashews into packaged products that are sold in India and internationally. Labourers are generally women working in dangerous conditions and although they do not benefit from price increases, the processing units provide steady work.

Keywords: cashew; Orissa; Odisha; India; food processing; workplace conditions

Background

The cashew industry in India is valued at US$800 million and according to a report in *The Guardian* newspaper it provides employment to approximately 1 million people directly in the processing of cashews and another 200,000 as growers (*The Guardian*, 2013).

Trimurti Cashew Industries is a processing unit that converts raw cashews into a refined food product. It has operated in Ranipetta village, in the coastal area of South Odisha in the Bay of Bengal, for more than 15 years. The unit was established at a cost of US$7,000, which was borrowed from the Small Industries Development Bank of India, a national-level bank responsible for promotion of small and medium enterprises in the country. The unit has the capacity to process 160,000 kg of cashews annually. The raw cashews are mostly procured from producers within a radius of 20 km, via local-level aggregators or agents. In 2013 the unit bought cashews at the rate of $1.20 per kilogram; 97 per cent of this goes to the producers and the balance to the agents.

For processing cashews the unit employs 40 labourers, almost all women, from Ranipetta and surrounding villages. They are employed for eight months every year to clean, break, roast and sort the raw nuts, and to package them as fine edible cashews to be sold via traders within India and to consumers in Europe. According to the Micro, Small and Medium Enterprises Development

http://dx.doi.org/10.3362/9781780448671.014

Institute (MSMEDI, 2013) an average of 10 per cent of the production of the cashew cluster in Odisha and Andhra Pradesh is exported, and current production in the country accounts for 45 per cent of global production.

In addition to the industry's significance at a global level, the cashew cluster in these states plays a crucial role in employment generation. Businesses such as Trimurti Cashew Industries provide market access for small-scale cashew producers of the region, and employment to large numbers of poor women from the region, although they are not necessarily perfect jobs. The processors also encourage other service providers to benefit from the value chain. This case study analyses the benefits that the industry generates for small-scale producers and women labourers. It also highlights the significance of the processing units in the industry.

From producer to processors – the cashew supply chain in South Odisha

South Odisha is a backward area, with very low socioeconomic indicators, and the region is also affected by violent extremists. In such a region a value chain that supports the livelihoods of a vulnerable population plays an important role. Figure 14.1 is an overview of the various actors involved in the chain and the specific functions carried out by them. Although it shows actors in various locations, this case study focuses on South Odisha.

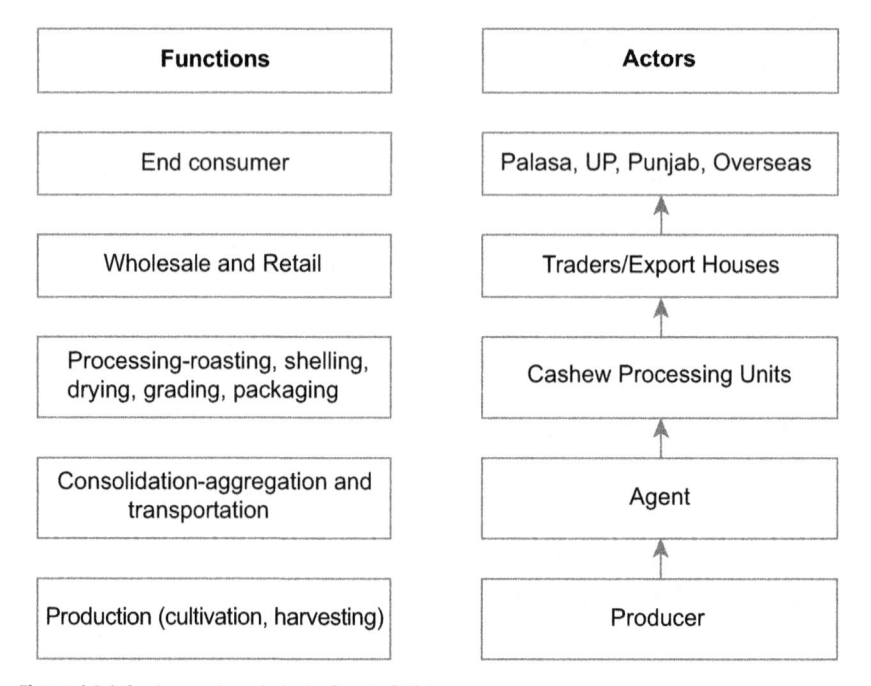

Functions	Actors
End consumer	Palasa, UP, Punjab, Overseas
Wholesale and Retail	Traders/Export Houses
Processing-roasting, shelling, drying, grading, packaging	Cashew Processing Units
Consolidation-aggregation and transportation	Agent
Production (cultivation, harvesting)	Producer

Figure 14.1 Cashew value chain in South Odisha

The story of small-scale producers in the value chain

Cashew growers in India are a critical part of the global value chain as they produce 60 per cent of the commodity consumed globally (*The Guardian*, 2013) According to the Directorate of Cashew Nut and Cocoa Development (DCCD, 2012), the top five states with sizeable cashew producing area in 2009–10 were Maharashtra, Andhra Pradesh, Odisha, Tamil Nadu and Karnataka. In terms of production, Maharashtra was much ahead of the others. It produced double the quantity of Andhra Pradesh. Odisha produced 13 per cent of India's cashew in 2009–10 and was third on the list.

Similar to any agricultural commodity, the value chain of cashew in South Odisha starts with producers or farmers who are mainly small in scale – the average land used for cashew plantation is around two acres – and engaged in cultivation and harvesting of cashew plantations with technical support from processors and other business development service providers. The production process in the country is varied and there are also many government and private plantations, but this study is confined to the smaller private farmers, processors and traders in Odisha.

Krushna is a typical small cashew producer who lives in Munsingh village in the Gajapati district of Odisha with his family. He inherited two acres of agricultural land from his parents. In 2013 he used 1 acre of land for cashew production, with around 70 trees. He spent $50 on fertilizer and medicines to ensure good-quality produce. Most of the labour that was required was from his own household. After the cashew season from March to June, Krushna harvested 300 kilos of cashew. Despite the volatility of the market, he was able to secure $1.17 per kilo of cashew. His gross income from cashew was $350 per acre of land in 2013 and his net income was $300. His income in 2012 was lower mainly due to lower prices. The income from cashew covered a significant part of the family's expenses in 2013. To meet his remaining expenditures, Krushna and his wife and children move to Hyderabad, a metropolitan city around 800 kilometres from his village, to work as unskilled labourers at building construction sites.

Another larger farmer in the same village has 14 acres of land with a total of nearly 1,000 mature high-yielding cashew trees. Each tree produces about 20 kilos of cashews in a year. Last season he sold these for about $1.10 a kilo, which earned him a total of almost Rs1.5 million, or about $23,000. He had to hire three labourers at Rs100 a day each for several weeks, but his net earnings from cashew were still substantial.

These two farmers are typical of the smaller and the larger cashew producers in the region. Gumma block in Gajapati district has approximately 5,000 families involved in cashew plantation and collection (MART, 2012). These households produce roughly 100 metric tonnes of cashew from 3,000 acres of land, and most of them are small producers. The small land holdings used for cashew restrict producers' ability to benefit from the lucrative and rising prices of cashew. Exports of cashews increased by 16 per cent in the nine-month period ending in December 2013 (*Economic Times*, 2014). There has also been a 24 per cent increase in value during this period, although this has mostly benefited the processors and exporters (*Economic Times*, 2014). Nevertheless, producers such as Krushna have also benefited from the price rises. The average price received by producers in 2012 was $1.08 per kilo, and this increased by 8 per cent in 2013. In addition to the price rise, there is a potential for small-scale producers to benefit more from the value chain by carrying out primary-level processing activities such as cleaning and grading. This however threatens the livelihoods of women who do this work at the numerous processing units.

Different types of aggregators in the value chain

The next level of actors in the chain is agents who act as aggregators, buying small quantities of raw cashew from several farmers and selling in bulk to the processors. They buy directly from the farmers and add their mark-up before selling to the processing units. The agents are mostly local village-level traders with sufficient capital to invest in buying and storing raw cashews. The processing units usually ask these agents to supply certain quantities of raw cashews, although sometimes the agents contact the processors after they have purchased the cashews. Cashew is a high-value seasonal crop, and there are many intermediaries who supply cashew from the villages to the processing units, and most processors have links to agents in convenient localities, from whom they source their raw material. The agents are also an important source of credit for the farmers.

Processing units

In the cashew industry, the processors are the main link of the value chain and the entire cashew economy revolves around them. According to the Indian Directorate of Cashew Nut and Cocoa Development 2005–06, there were 3,799 cashew processing units in the country in 2005 (DCCD, 2013). Based on their capacity to process a certain number of 80 kg bags, they are categorized as big, medium and small processors. The big processors are able to process more than 25 bags per day, medium between 20 and 25 bags, and small ones handle a maximum of 20 bags per day. This classification helps to understand the kind of capital investment and technology used for processing of cashews. The bigger ones tend to have more productive and safer technology, whereas the smaller ones are often labour intensive and risky.

The geographical spread of the processing units shows a varied degree of concentration. About 75 per cent are in the states of Maharashtra, Kerala and Karnataka. Odisha and Andhra Pradesh have roughly 10 per cent of the processing units (DCCD, 2013). Their combined capacity is roughly

12 per cent of the country's overall capacity. The average capacity of a unit in these two states is roughly a third of that of a unit in Kerala or Karnataka. Cashews processed in these two states are mostly produced locally. The stated investment in units in Odisha and Andhra Pradesh varies from around $2,000 to over $70,000 (DCCD, 2013), and the number of employees ranges from 20 to 60, most of whom are women. The raw material procured by the processing units is roasted, shelled, dried, graded and packed. Almost all the units use the conventional drum-roasting method for cashew nut extraction.

Traders

Once the product is ready to eat it is supplied to traders across the country, including export houses for supply to international consumers and local retailers within India.

Prices in the value chain

The price of cashew at different levels of the value chain in the cluster is based on the expenditure incurred by the specific actor and overall demand. Figure 14.2 shows the price per kilo of cashew from the producer to the retailer. It is evident from the figure that the maximum value is added by the processing unit. The percentage change in the price does not of course mean that the actor earns that much profit.

The average cost of raw material per 100 kilos in 2013 was $100. Another $41.67 was spent on variable items related to finishing of the raw material. The breakdown of the cost is 71 per cent for raw material, 27 per cent for labour, and 2 per cent for transport, packaging and promotion. The detailed cost of processing per 100 kilos of cashew is shown in Table 14.1.

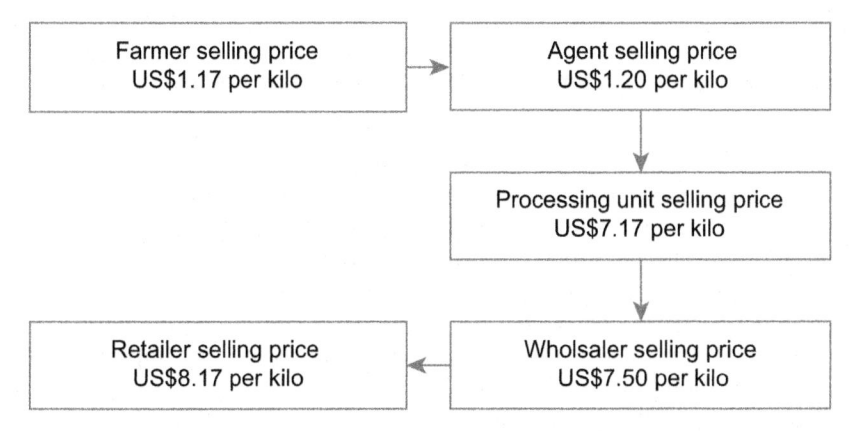

```
┌─────────────────────────┐      ┌─────────────────────────┐
│ Farmer selling price    │ ───► │ Agent selling price     │
│ US$1.17 per kilo        │      │ US$1.20 per kilo        │
└─────────────────────────┘      └─────────────────────────┘
                                              │
                                              ▼
                                 ┌─────────────────────────────┐
                                 │ Processing unit selling price│
                                 │ US$7.17 per kilo            │
                                 └─────────────────────────────┘
                                              │
                                              ▼
┌─────────────────────────┐      ┌─────────────────────────┐
│ Retailer selling price  │ ◄─── │ Wholsaler selling price │
│ US$8.17 per kilo        │      │ US$7.50 per kilo        │
└─────────────────────────┘      └─────────────────────────┘
```

Figure 14.2 Selling price of cashew at different levels of the chain

Table 14.1 Average cost of processing 100 kilos of cashew in a processing unit in Parlekhamundi

Particulars	US$
Raw material	100.00
Transport	0.83
Labour for handling raw supplies	0.50
Labour for roasting	1.67
Labour for breaking the shells and extracting kernels	19.17
Labour for sorting and cleaning	16.67
Packaging	0.83
Transport for selling	0.83
Promotion	0.67
Labour for handling finished product	0.50
Total cost	**141.67**

When processed, 100 kilos of raw cashew produces 30 kilos of edible cashew nuts of different qualities. The three best qualities of cashews are known as 180 grade, 210 grade and 240 grade. Each differs in size and quality and hence in price. At the time of our study, the 180 grade was priced at $6.67 per kilo, 210 at $5.83 and 240 at $5.00. About two thirds of the final product can usually be processed to grade 180. The remaining third is equally distributed between 210 and 240 qualities. Based on the selling price offered for the various qualities of cashew, the gross sales after processing 100 kilos of raw cashews was $187.50, which gives a net margin of 32 per cent. This margin can differ depending upon the capacity and technology used for processing.

Opportunity and risks for the workers employed by processing units

The employment generated by the cashew industry is substantial, as already stated; most of the workers are from rural areas and are often poor. The information collected for this case study suggested that most of the workers live below the poverty line. Around 90 per cent of the workers are women and the data collected show the average annual income of a woman worker is roughly $350. This income is earned in eight months and is roughly a fifth of a typical household's total annual expenditure. During the interviews many women said that their earnings from cashew processing had improved their status in relation to the men in their families.

In addition to regular cash income, the workers obtain few other benefits. The number of units that provide free meals at work, access to health services, paid annual leave or severance pay in case of illness or work accidents, is negligible. In a few cases, a trade union has been set up and a nursery constructed, with a clean, sheltered area where mothers can arrange someone to look after their babies, but no food is provided and there are no trained child carers, such as were apparently available in the older government-owned factories. Only the relatively large factories that have significant investment in state-of-the-art technology to process raw cashews have even these basic childcare facilities.

The women who work at the smaller factories said that their earnings were at least 10 per cent lower than their counterparts in bigger factories. Cashew processing also involves working conditions that expose workers to many different risks. Some women who earn as little as fifty cents a day suffer permanent damage to their hands from the corrosive oil that emerges when the nuts are broken. The most significant difficulty in processing cashew nuts is that the shell contains this caustic oil, which can burn the skin and produce noxious fumes when heated. There is a particularly high risk of hand injury when the workers use hand-held hammers to separate the nut from the shell. In the smaller factories the raw nuts are steamed and are then cooled and cut with a hand and foot pedal-operated machine. The workers are given oil to cover their hands, but this provides only limited protection. The workers can also wear gloves, but these wear out quickly and in any case, are not favoured by workers who are paid on a piece rate basis, since the gloves reduce the quantities of nuts they can process. Some of the women said that they suffered from pain in their hands and their shoulders, and there were also some women who had arthritis and diabetes.

In spite of these problems, however, many of the workers were happy with their jobs. Promdini Mali, for example, is a widow with two children, a daughter and a son. Both of them are studying in the nearby village government primary school. This would have only been a dream for her if she had not obtained a job as a worker at the Shri Krishna Cashew Industry, one of several cashew-processing factories in the area. She is mostly involved in roasting, shelling and packaging, and she is paid on a piece rate system at Rs150 per 10 kilos of cashews processed in a day. In 2013 she earned about Rs35,000 or $600 for eight months' work. Her responses to the Progress out of Poverty Index (PPI) questionnaire as to her situation in 2008, two years after her husband's death, suggested that the likelihood that she and her children were living in poverty at that time was over 75 per cent. The likelihood of the family still living below the poverty line at the time of our meeting in December 2013 had dropped to 41 per cent. Without this job it is hard to imagine that she and her children would have moved out of poverty.

Growing pressure from the international development community

The suffering of women in the industry has recently reached the attention of the world community. According to a study by Action Aid (2011), only 1 per cent of the price paid for cashew at a UK supermarket goes to the worker.

The producers get 15 per cent of the value, while the remaining 84 per cent is shared with other players in the international supply chain, including the UK retailers, who get 45 per cent of the retail price. The study described how retailers in the UK bring pressure on their suppliers to reduce their prices, which makes them reduce their labour costs by paying lower wages. The study highlights the unsafe and unhealthy working conditions. This growing awareness has put pressure on governments and private businesses to take action to make the work safer for employees, especially for women workers. Although there are certain codes of conduct that are designed to improve working conditions for the poor who work in the cashew nut industry, there is still a need for stricter legislation.

References

Action Aid (2011) *Who Pays? How British Supermarkets are Keeping Women Workers in Poverty*, London: Action Aid International UK.

DCCD (Directorate of Cashew Nut and Cocoa Development) (2012) *Annual Report*, Kochi: DCCD.

DCCD (2013) *Annual Report*, Kochi: DCCD.

Economic Times (2014) 'Cashew exports touch all-time high', *Economic Times*, 24 March.

The Guardian (2013) 'Cashew nut workers suffer appalling conditions', *The Guardian*, 2 November.

MART (2012) 'Scouting and developing livelihoods opportunities in Orissa', New Delhi: MART.

MSMEDI (2013) *MSMEDI Annual Report 2013*, Mumbai: MSMEDI.

Acknowledgements

Data for this case study were obtained by the following MBA students from Centurion University of Technology and Management, Paralekhamundi campus: Sandeep Panigrahy, Bhaginath Subudi, Tulusi Ram, Amit Singh, Ranjot Kumar Nayak, Rakesh Kumar Das, Pratyusha Kumar, Mohanta, Sarat Chandra Gouda, Tejaswini Mohapatra, Dipika Kumari Pradhan, Mahesh Kumar Nayak, Siva Prasad Choudhary, Nalinikant Choudhary, Kunal Digal, Simalchal Mohanty, Binupakhya Patnaik, Punit Kumar Odhya, Kamaol Kant Biswal, Mahesh Tripathy, Nanasingh Nath Kara, Neha Singh and Padma Baruda. Their faculty supervisor was Dr Umakanta Nayak. Their assistance is gratefully acknowledged.

About the author

Kulranjan Kujur studied public affairs at Cornell University, USA. He is passionate about addressing poverty issues through the development and involvement of the private sector.

CHAPTER 15
Palm oil in Peru – small-scale farmers succeed where plantations failed

Rafael Meza

Abstract

This case describes the value chain of palm oil for animal feed in the Amazon region, more specifically in the Loreto and Ucayali provinces of Peru. The product is used by local poultry and pig producers, substituting animal feed imported from outside the region. The large-scale technology typical for palm oil production and processing was not successful in the Amazon but a local company, Agro Selva Ltd, benefited from the earlier investments made and revived the abandoned production sites by engaging small-scale farmers and installing small-scale processing technology.

Keywords: palm oil; animal feed; import substitution; Peru; smallholders

History – the failure of a large-scale approach

In 2005 OLAMSA (Amazonicas Oil Company Anonymous) started establishing a palm oil plantation in the Peruvian Amazon. It convinced the regional governments of Ucayali and Loreto to provide support. With the help of the local mayor a nursery was set up in Pampa Hermosa (Ucayali province) in 2007, aiming at growing seedlings for a pilot area of 500 hectares. Harvest would start in 2010, three years after establishing the plantation. The regional government of Loreto supported the initiative by providing subsidized credit to cover planting and maintenance costs. In the original set-up the company and the government would jointly finance the construction of a processing plant.

Difficulties quickly arose right after the start of the project. The context of the Amazon appeared very challenging for large-scale palm production, particularly the unfavourable climate and the high transport costs, resulting in expensive inputs and equipment which, hampered the progress of the project. Soon it became clear that the large production area required to make investment in a processing plant feasible could not be realized. Also, the cost of the processing plant was much higher than projected due to high transportation and construction costs – a typical feature of the Amazon. What did not help either was that the commitments made by regional and local governments appeared less solid than expected. In this context, the interest of the community in the initiative also waned.

http://dx.doi.org/10.3362/9781780448671.015

The future potential for large-scale palm production in the Peruvian Amazon is not favourable at all. The agricultural area in the Peruvian Amazon is relatively small and it is estimated that only 2 per cent of the area is suitable for palm tree plantations. The suitable areas for agriculture are scattered, hampering large-scale exploitation. In the Amazon an oil palm plantation is expected to be productive for about ten years only, due to the rapid depletion of the soil. Developing palm plantations is expensive and therefore requires long-term returns, covering much longer than ten productive years.

Legal issues offer challenges too. The major part of the Peruvian Amazon is forest area and in Peruvian law the only legal form of access and use is a forest concession, not private ownership. The forest concession describes the conditions for exploitation, reforestation and conservation. Oil palm trees are not qualified as reforestation since it is an exotic species and grown as a monoculture, which is not allowed.

There are environmental limitations too. Road construction and demand for labour generate a process of occupation and colonization, which has a large impact on the fragile Amazon ecosystem. Experience has shown that such processes often trigger negative impacts such as land invasion by farmers, illegal logging, hunting and land appropriation by timber operators.

Finally there are social constraints. The local people traditionally cultivate between 5 and 30 hectares. Larger areas require more labour, which conflicts with other traditional activities. The population is engaged in diversified economic activities such as hunting, fishing, logging and horticulture. Palm tree cultivation is a complementary activity that will not replace other activities.

Agro Selva Peru tries a different approach

In this context Agro Selva Peru stepped in at the end of 2009. In an attempt to rescue the established plantation of around 240 hectares it mobilized 25 producers to do the daily maintenance work. The owner of Agro Selva had been working for many years in training palm oil producers, both inside and outside Peru, and concluded that the large-scale approach could never work for the Amazon. He decided to try out a small-scale approach. At the same time he also saw great potential in supplying the growing local poultry and pig production industry, responding to the growing demand for meat and eggs in the region. He also realized that the existing abandoned palm plantations in the area provided a resource base for this industry, which could be exploited at a limited cost.

New small-scale technology was already available. Agro Selva had developed appropriate technology for exploiting palm in the region, a simple processing machine handling small volumes. The standard bulk processing technology had proved to be unfeasible while the small-scale technology made exploitation of the available, previously abandoned palm tree plantations possible and viable.

The owner had gained experience in palm oil processing and production in other regions, within and outside Peru. He had organized many training programmes and provided technical assistance to palm producers. Reflecting on the large-scale business model so typical for palm oil, he concluded that in the Peruvian Amazon production volumes would never suffice for a medium or large processing plant.

Agro Selva Peru SRL is a small company with a yearly turnover of approximately US$440,000. The company's owner provided the financial capital for the company; there are no investors from outside.

Processing is organized around strict procedures to avoid contamination with microorganisms. The pasteurization process follows a production protocol that includes hygiene, water filtering, appropriate packaging, the use of additives and so on. The processing plant installed by the company in Pampa Hermosa involves simple technology and is relatively small, with a processing capacity of 2.5 tonnes of oil per day, or one tonne per shift of 8–10 hours. The plant is powered by a small diesel engine of 15 horse power; the village does not have electricity. The plant required an investment of approximately $80,000.

The strategy the company used to increase supply was to pay a premium price of 65 cents per tonne, compared to other purchasers and to provide technical assistance. The producers were trained to perform tasks that require some expertise and practise such as pruning, harvesting, fertilization and post-harvest handling. In total 88 smallholders have been trained. Providing a price premium for larger volumes and signing temporary supply contracts were other ways in which the company attempted to strengthen its linkages with its suppliers.

In 2011 and 2012 the company was not profitable, mainly because it was investing heavily in infrastructure, mainly processing equipment, and paying producers extra to motivate them to harvest the palm fruits. In mid-2012 the international price of palm oil dropped suddenly, putting local prices under pressure. The lower prices had a serious impact on Agro Selva's financial situation; something the company is still trying to recover from. The company is still not profitable, and immediate improvements are needed. However, the demand for palm oil is growing, and the regional market for value-added products such as soaps, palm oils and stearin will increase significantly, so the future seems bright for Agro Selva. The company also expects to increase its profitability by selling its by-products, particularly waste water, which is rich in nutrients and can be used for animal feed, compost and organic fertilizer.

The palm oil value chain

The first stage of the value chain involves 25 small-scale farmers managing an area of 300 hectares of oil palm. On average, each farmer manages 12 hectares of oil palm. These smallholders can be characterized as subsistence farmers since they mainly produce for their own consumption; the little income they earn is just enough to meet their minimum basic needs. In their area a hectare of fallow land costs as little as between $80 and $150, depending on factors such as availability, quality and distance to the village. The investment needed to establish a hectare of oil palms is usually between $2,000 and $2,500, but the new suppliers to Agro Selva did not have to meet this cost because the investment had already been made by the earlier unsuccessful business.

According to a recent forest census, a palm tree plantation has an average of about 140 palm plants per hectare. The production of palm oil fluctuates between 960 and 4,800 kilos, depending on weather conditions and management.

After the cultivation and harvesting of the fruit, the second stage in the chain is the transformation of the fruit into palm oil by Agro Selva. The company's small-scale processing plant is located close to the production areas. The plant transforms the fruit into palm oil by threshing the fruit bunches, cooking them, pressing and refining the oil through centrifuging and heating. To produce 1 kilo of oil, 5 kilos of palm fruits are needed. The refined oil is packed in drums of 200 kilos for transport and sale. It takes two days to bring the product by river to the buyers in Pucallpa, and four days to Iquitos. The business adds between 45 and 50 per cent of the value generated in the value chain.

The third link in the chain is a company called Q&C that buys the palm oil from Agro Selva and manages its sale and distribution to the final users. In order to better serve their end markets, Agro Selva contracted this firm to be their sales agent in Iquitos and Puccalpa. Agro Selva purely focuses on organizing the production and processing of the palm oil while Q&C handles the marketing and sales.

Value Chain

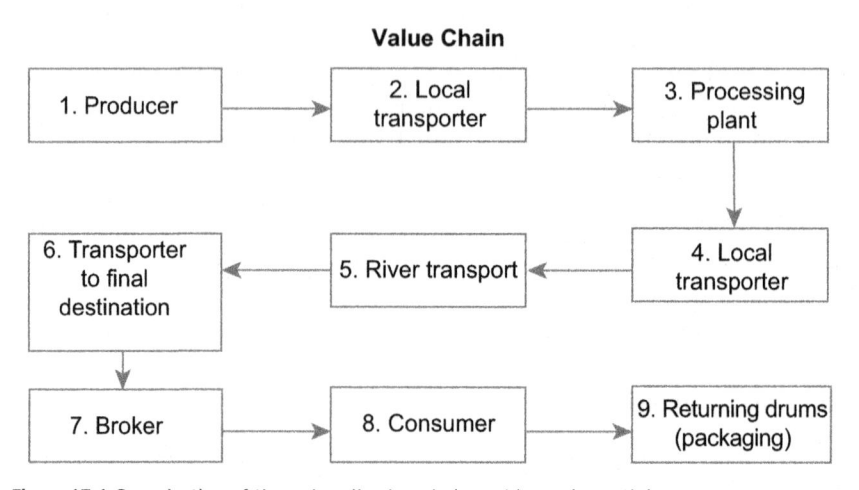

1. Producer → 2. Local transporter → 3. Processing plant → 4. Local transporter → 5. River transport → 6. Transporter to final destination → 7. Broker → 8. Consumer → 9. Returning drums (packaging)

Figure 15.1 Organization of the palm oil value chain and its main participants

The fourth link involves poultry farms, and to a lesser extent pig farms, located in Iquitos and Pucallpa, the two major towns in the Peruvian Amazon region. These farms buy the palm oil to supplement the diet of their animals. Due to increasing local demand for meat and eggs, this industry has been growing steadily over recent years.

Some business services providers operate along the chain, facilitating operations. Agro Selva itself provides assistance to smallholders on how to manage their trees. Other providers serving the chain include transporters (mainly river transport), suppliers of second-hand oil drums in which the palm oil is packed, laboratories that carry out quality analysis and suppliers of farm inputs. The district municipalities promote palm oil exploitation and oversee the operations in their districts.

The end market

The palm oil provides additional energy and has become a vital component of the animal feed used by poultry and pig farms in Iquitos and Puccalpa. Agro Selva stresses the high quality standards of its product, a result of applying strict protocols at each stage of the production process. The company emphasizes that its product has comparative advantages over other types of animal feed, such as the commonly used cotton oil, which has the same calorific value but is sold at a higher price.

Poultry and pig production is a thriving and growing industry in the region. Oils and fats are used as a feed supplement at a rate of 3–6 per cent of the animal feed mixture. In the region an average of 80,000 chickens per day are consumed. This amounts to some 144 tonnes of poultry meat, which requires about 260 tonnes of poultry feed per day. This translates into a daily demand for 8 tonnes of crude palm oil. No similar data are available on the volume of pig farming, but pork consumption is growing in the region and the demand will surely rise in future.

An official of the Regional Directorate of Industry stated that the palm oil value chain has the potential to become an important economic activity in the region. In 2010 the demand for crude palm oil as supplementary animal feed was estimated to be between 30 and 40 tonnes per month in Pucallpa and between 50 and 60 tonnes per month in Iquitos. At present Agro Selva sells approximately 14 tonnes of crude palm oil per month in Pucallpa, thus satisfying between 35 to 45 per cent of the market, while in Iquitos they sell between 8 to 10 tonnes per month, which is between 15 to 20 per cent of the market.

Taking into account all the costs of inputs, labour and services, covering all the stages from cultivation, supply to Agro Selva, processing and marketing, the total value added in the chain in 2011 was $180,000. In 2012 this had increased to $250,000 (see Table 15.1).

Table 15.1 Production costs for 1 tonne of crude palm oil

Description	US$	% value chain
Producer/farmer	270.0	37.5
Processing costs	289.0	40.0
Loading	18.5	2.6
River transport	46.0	6.4
Unloading	15.0	2.1
Market transport	12.5	1.8
Broker	36.0	5.0
Administrative costs	18.0	2.5
Return on package	10.8	1.5
Management	5.0	0.6
Total	**720.8**	**100**

The production costs presented in Table 15.1 are average figures over the last three years. The sale price fluctuates with changes in competition and demand at the level of the local and world market. The OLAMSA Company, which is a large palm oil producer in Peru, sells a tonne of crude palm oil at $775.00.

The processing cost is relatively high, accounting for 40 per cent of the final cost of the end product. This may decrease when Agro Selva realizes improvements in handling and processing the fruits and by generating additional income through the sales of by-products.

Benefits to the value chain

Twenty-five young male and female farmers supply palm fruits to the company. The 12 permanent staff working at the processing plant are local farmers who have been trained by Agro Selva. Taking an average of six members per family, in total about 200 people are direct beneficiaries of the initiative. The indirect beneficiaries include about 120 families who are involved in providing labour, transport and input supply.

Income data obtained from 25 producers supplying fruit to Agro Selva in 2011 and 2012 show that the average yearly income generated by palm oil was around $2,600 in 2011, which increased to $3,300 in 2012. One of the more successful producers was able to earn more than $9,800 in 2011 and a similar amount in 2012. At the lower end, the smallest supplier obtained a little over $100 in 2011 and almost $200 in 2012. In other words, there is a large variation in the income generated from harvesting palm fruits among the farmers involved. Additional field research is needed to find out what are the reasons behind this.

The income earned by the small oil palm producers in Pampa Hermosa is stable and significant; and it complements their other traditional activities. So long as they do the minimum of cultivation it is possible to generate about $1,600 per year from the typical 5-hectare production area. If the farmer manages the palm trees more thoroughly, this can increase up to $5,850 a year. This is an attractive additional source of income, certainly since the opportunity cost of the producers' labour is close to zero in this area of limited economic opportunities. It is sufficient to encourage other people to become involved in palm oil fruit cultivation. Compared to other income-generating activities, managing a palm plantation is an ongoing activity that can generate a steady income, providing cash throughout the year.

Impact on poverty

Twenty farmers were interviewed to gather information on how oil palm has improved their lives. They were chosen from the group of palm fruit producers that were available during the field research. The Progress out of Poverty Index (PPI) questionnaire for Peru was used, but this created some difficulties. The Amazon is a totally different environment than the rest of Peru, and therefore some of the questions, typically those about housing and possessions, were out of context. Other items, not included in the PPI question list, are more relevant for measuring relative wealth in the Amazon, such as owning an outboard engine or a motorized canoe, items that have little relevance in other regions.

Before the establishment of the palm tree value chain, the subsistence farmers were living in extreme poverty. Their income per capita was far below $1 per day; health conditions and life in general were precarious for almost all the farmers. Community life and social cohesion were meagre considering the few social activities that were organized (such as religious festivals, political rallies or community parties). The exploitation of palm oil generated additional income for the 20 farmers interviewed. To date the monthly income earned from selling palm fruits is on average $180 for each family. This enables them to meet some of their basic needs, particularly improving their diets and general living conditions, mainly housing. For example, two thirds of the surveyed farmers reported that they had replaced the earth floors in their houses with cement.

The respondents mentioned that thanks to the income from selling palm fruit, they bought things like an outboard engine, a *peque peque* (a motorized canoe, a typical form of transport in the Amazon), a chainsaw, a power generator, musical equipment, a telephone, a refrigerator, a television set, a corrugated metal roof, or a motorcycle or motorized rickshaw. They also mentioned they had travelled more often, and organized more parties. About a fifth of the respondents managed to buy a freezer or kerosene refrigerator and used it to sell ice and ice creams. Of the 20 farmers, 19 have a television set at home now and 2 have more than one set.

One palm fruit producer described his progress out of poverty as follows: 'With the sale of the palm fruits I was able to travel to Pucallpa several times and bought my own *peque peque*, so yes, I would say I'm making progress'. Another producer, Rusber Sangama said: 'I'm growing with the sale of palm oil; I bought my own generator so now I have light in the evening'. Producer Hilda Cushuchinari said: 'with my harvest I have bought clothes for the entire family and a cell phone'.

This value chain, small as it is, has had a large impact on those rural people who are involved in it, mainly since it generates a steady income flow for them. This makes a big difference from before when they depended on the irregular small income they could get from hunting, fishing and small-scale farming.

The future

Responding to the increasing demand, Agro Selva is expanding its supply base and engaging more farmers. Based on the experience it has gained up to now, new facilities in other locations can be opened, thereby boosting palm cultivation in the entire region. Agro Selva's owner foresees that he can roll out the model to many more communities, thereby integrating these more into the regional economy.

There are good reasons for his belief that the business model can be substantially extended within the Amazon region. There is a continuing growing local demand for oils and fats to supplement the diets of poultry and pigs. It is estimated that the demand for crude palm oil can increase to up to 200 tonnes per month. One of the main buyers of the product, Fernando Zelada, confirms this:

> I am willing to buy larger volumes. I want to use more palm oil since traditional animal feed is getting more expensive while the competition puts a pressure on my sales price. An additional benefit is that consumers prefer the dark coloured eggs, which chickens fed on palm oil tend to lay.

Also, transport difficulties and the high freight rates for feed supplements from outside the Amazon region give locally produced palm oil a comparative advantage. At the production side, the lack of other sources of income for the small-scale producers means that they are eager to get involved.

Three other producers have followed Agro Selva's successful example and have started to process palm oil for themselves, using traditional methods. There is a risk that the low quality of their product will jeopardize the reputation of the whole value chain. These producers operate informally and can operate at a much lower cost, and can therefore sell their oil at a lower price than Agro Selva. In order to deal with this problem Agro Selva offered to allow them to use its plant in return for a small service fee, and to jointly market the product to save on transport and marketing costs. Agro Selva's owner also indicated he would be willing to allow these informal competitors to be partners in his company if the other options were not acceptable. At the time of writing, these informal producers were considering whether to continue on their own or to work with Agro Selva. The company's owner is very anxious to find a solution because he wants as much supply as he can to serve the growing market and is afraid that the market may collapse if lower quality palm oil enters the market.

About the author

Rafael Meza is a lawyer, specializing in legal access to natural resources. His experience combines natural resource management and local economic development in the Peruvian Amazon.

CHAPTER 16
Organic turmeric from eastern India – healthy spice and healthy earnings

Niraj Kumar

Abstract

This case study describes a small local enterprise-led inclusive value chain. The chain strengthens existing traditional farm practices followed for centuries by tribal communities in a heavily forested area of India, but also improves the socioeconomic well-being of farmers without affecting the environment. Government agencies and community-based institutions are crucial for the chain's success. The case shows that a well-executed business can benefit small traditional farmers along with other players in a value chain.

Keywords: turmeric; India; Odisha; organic; tribal

Biosourcing (Biosourcing.com) is a small company dealing with ayurvedic and herbal supplements and organic products. Its headquarters are in Bhubaneswar, the capital of Odisha state in India. The company believes in Ayurveda, the ancient Indian system of medicine based on natural holistic food, and aims to make its natural herbal products available throughout the world. The company has customers in 12 countries in 5 continents, works with about 1,500 farmers in the cultivation of organic products, and has an annual turnover of more than Rs50 million or about US$1 million.

It is not easy to estimate the total area in India that is under organic cultivation. There are a number of different certifying agencies, and there are of course large areas of forest that are uncertified but have never been anything other than organic. A 2005 paper (Bhattacharya and Chakraborty, 2005) stated that 76,000 hectares of land were certified organic, along with 2.4 million acres of forest that was effectively organic, and that total exports of organic produce in 2004 amounted to 6,472 tonnes. A publication by the United Nations Food and Agriculture Organization (FAO, 2004) stated that total organic production in 2004 was 14,000 tonnes, of which 12,000 tonnes were exported. Another publication (Charyulu and Biswas, 2011) mentions a figure of 37,500 tonnes of exports in 2007–08, worth $100 million. Whatever the numbers, it is clear that the present production is large, and that there is great potential for expansion in the future.

http://dx.doi.org/10.3362/9781780448671.016

The data for the state of Odisha are also confusing; an article in the *New Indian Express* on 13 August 2014 stated that 69,055 hectares in the state are certified, but the same article also stated that 18,500 hectares are under organic cultivation. Regardless of such confusion, there is certainly large potential, particularly in the forested hill areas that are largely inhabited by so-called 'tribal' people or 'tribals'. This is an umbrella term for the minority aboriginal population of India, and like the so-called 'scheduled castes' (SC) they are mentioned in the Indian constitution as 'scheduled tribes' (ST).

Organic product sourcing by Biosourcing in India

Biosourcing tries to meet its orders primarily from Odisha. If this is not possible from its own local partner-farmers it uses well-established certified suppliers from the state and suppliers outside the state with which the company has contacts. The company works with 1,500 farmers, who are mainly tribals and from tribal-dominated districts such as Kandhamal, Koraput and Bolangir. The company has its own laboratory for quality testing and control, its own processing unit, and has also opened an organic retail store in Bhubaneswar, called Nature-Natural. This shop in the heart of city not only sells packed organic and natural products but also offers the services of an ayurvedic doctor who prescribes appropriate medicine and treatment.

The company helps farmers by encouraging community-based farming and direct procurement from farmers' institutions, by removing middlemen, and by rewarding organic farmers with timely and better payments for their produce. At the same time the company increases the value of its products for its customers by providing the best possible quality products at competitive prices which meet the standards of the internationally accepted HACCP (Hazard Analysis and Critical Control Points) and GMP (Good Management Practices) certifications. The company supplies herbs and spices in raw form, as powder, blends, extract, oil, alkaloids and formulations in bulk to customers from countries such as Australia, Canada, Denmark, France, Lebanon, South Africa, United Kingdom and United States.

The company maintains complete transparency in product information and quality control, and shares all the reports of their laboratory analyses with customers. If required, customers can be provided with details of farmers' backgrounds, fields and cultivation practices. The company also facilitates customer visits to farmers.

Although the company can provide a variety of organic products, it regularly receives orders for turmeric (*Curcuma longa*) powder, arrowroot (*Maranta arundinacea*) powder, ginger (*Zingiber officinalis*) powder, cumin (*Cuminum cyminum*) powder, coriander (*Coriandrum sativum*) powder, fenugreek (*Trigonella foenum-graecum*) powder, black pepper (*Piper nigrum*), chilli (*Capsicum annum*) flakes and powder, triphala powder (a combination of three fruits: *Emblica officinalis, Terminalia bellirica* and *Terminalia chebula*) and amla (*Phyllanthus emblica*) powder. As organic turmeric is one of the most important products for which the company is known, constituting 20 per cent of its total business, it is the central subject of this case study. Turmeric, a spice, comes from the root of a green plant of the *Zingiberaceae* family. It is used as a condiment, dye, drug and cosmetic in addition to its use in religious ceremonies in India.

Biosourcing procures organic products mainly from districts such as Koraput, Bolangir, Kalahandi, Kandhamal and Rourkela. However, in the case of turmeric, their entire sourcing is from Kandhamal (see Figure 16.1). The entire district lies at a high altitude and is made up of hilly ranges and narrow valleys. More than half the population are scheduled tribes and come from an original tribal race known as Kondhas. Almost two thirds of the land is covered with dense forest and mountains, and villages are scattered along the hillsides and in the valleys. Tribal people such as the Kondha, Kui and Kutia inhabit these natural forests. They have always practised organic cultivation and depend on forest products for their survival.

Turmeric production in Odisha

Turmeric has for many years been an integral constituent of traditional cultivation in the area. In the district of Kandhamal more than 60,000 farmers produce about 12,000 tonnes of dried turmeric every year from about 15,000 hectares of land. Kandhamal turmeric is said to have very high curcumin content, which is the main active ingredient of turmeric, and is famous worldwide for its colour, texture, aroma, flavour and long shelf life. There is a folk story in the area telling how the tribals would sacrifice a male child and offer his blood to the field so that the turmeric became more red. There are many private and government-promoted agencies such as OMFED (Orissa State Cooperative Milk Producers' Federation Limited) and KASAM (Kandhamal Apex Spice Association for Marketing) involved in promoting organic turmeric in the district.

It is vital to select the correct collection areas in the forests and the right farmers. The production areas, producers, production process and post-harvest activities must be certified by a certifying agency. It is the responsibility of the company to get the areas certified. The company follows a set procedure to select and register farmers as organic. The selected farmers are provided with cultivation guidance, field-level training and are encouraged to follow the guidelines. Certification is an ongoing process and the company keeps the certifying agency engaged to maintain the validity of the certificates crop after crop. Biosourcing spends about Rs1 million every year to keep its partner-farmers and their land 'certified'.

The company has also developed a minimum standard procedure for collection from the wild forest area, which includes registration and training of collectors, surprise visits to collection sites to check whether the proper harvesting process is being followed, audit of post-harvest methods, temporary storage and grading. These collectors are independent farmers and are members of a village-based forest protection committee. After ensuring that all the minimum standard procedures have been followed by each collector, the company procures the collectors' products.

Biosourcing prefers to procure its products from local community-based institutions. By working with these institutions they can give more negotiating power to farmers but can also more easily train the villagers in sustainable farming and harvesting and monitor their cultivation and collection activities. This is crucial in the business of organic products. Biosourcing also arranges loans for needy groups to start or support new value-addition activities. Further, community-based institutions help to forecast the quantity of products that will be available and to maintain a uniform high quality of procured material. The company pays the community institutions direct. Of Biosourcing's total business, 70 per cent comes from products grown by the farmers in Odisha. For turmeric procurement the company uses two major community-based institutions.

One type of institution is the Van Samrakshyana Samiti (VSS), or Forest Protection Committees, one of which has been formed by the state Forest Department in every village with the objective of jointly managing the forests. The initiative, known as Joint Forest Management (JFM), allows community-based institutions such as VSS to make decisions about the management of the forest and to share the benefits. Biosourcing directly engages representatives of VSS to decide the quantity, quality, cultivation practices and price of the produce.

The other type of community institution are the self-help groups (SHGs), which have been promoted by government and banks to help their members to pool their own money and community resources to meet their individual or the group's requirements. For Biosourcing, dealing with the representatives of VSSs and SHGs is easier as they do not need to engage individual farmers to ensure that all the guidelines regarding cultivation and post-harvest management are followed. As the company makes it payments to the SHGs, it also avoids the cumbersome process of distributing money to individual members.

The government has played a crucial role in the formation of community-based institutions such as VSSs and SHGs. There have been consistent efforts by the district administration to promote the cultivation and marketing of organic turmeric and to ensure better returns to farmers. The government's District Rural Development Agency (DRDA) plays a direct role in the promotion of crops in the area. Their local offices have information about each farmer, their crops and their links to markets in the same block. Any organization that has a business interest in organic products can contact the nearest DRDA office; this not only makes it easier to find a potential cluster of farmers but also reduces competition between community organizations.

Good quality and 'absolutely natural' turmeric makes Kandhamal one of the preferred sources of turmeric for both national and international buyers. In 2011, 2012 and 2013 Biosourcing on average bought 32–35 tonnes of dried turmeric from farmers in Kandmahal every year, and sold it to foreign buyers.

Biosourcing's turmeric value chain

Farmers either grow turmeric on their own land or they collect it from the forest. In both cases, the area where turmeric is grown has to be certified organic. The company either provides planting materials to the farmers or approves what they already have. Most farmers use their own planting materials, which they keep from previous years' crops. At every stage, company representatives monitor the cultivation practices, and whenever required they provide on-field guidance to the farmers. Farmers are also trained on harvesting and post-harvest practices including cleaning and drying, temporary storage and grading.

The company has two representatives, an agriculture officer (AO) and a forestry officer (FO), who work continuously with each farmers' institution. Ayurvedic experts supervise harvesting and post-harvesting activities to ensure that harvesting is sustainable and the products do not lose their characteristics during the post-harvest operations. After the crop has been harvested, the farmers clean, dry and store it. Special facilities have been created for the storage of organic products. Biosourcing inspects the stored product for conformity to the standards of the European Commission (EC) and the United States Department of Agriculture's National Organic Program (NOP). If the products are found to be suitable they are graded according to the official grading manual. The farmers are aware of the finer details of storage and grading. Loading and transportation are the company's responsibility. Every farmer gets a lot number along with the quantity and date of receipt for the product and it is then transported to the company's warehouse for further processing. Payment is made within a week of receipt of produce at the company's warehouses. All payments are made to the farmers' institutions and it is their responsibility to distribute the money among their members. A total of 460 farmers are growing turmeric for Biosourcing. Although the quantity supplied by individual farmers varies, on average each farmer supplies 250–300 kilos to Biosourcing each year.

If the farmers have more produce than Biosourcing is ready to buy, although this rarely happens with turmeric, or if farmers are not interested in selling to Biosourcing, they are free to sell in the open market. However, the market price is invariably lower than the price the company offers to the farmers. In 2013, on average, Biosourcing paid between Rs60 and Rs65 a kilo for turmeric, giving the farmers a profit of about Rs30. The open market price was around Rs40–45 a kilo.

The warehouse, on receiving the products, takes a sample of each lot and sends it to the company's laboratory for testing. If the laboratory approves the sample, it goes for further processing, which includes cleaning, drying, polishing, grading, grinding, storage and final packaging. Each batch of the final product is given a product code and information about its quantity and the date of processing. If the batch fails the laboratory test, it is sent to a non-organic store. The company then gets in touch with the farmer and takes

166 COMMERCIAL AND INCLUSIVE VALUE CHAINS

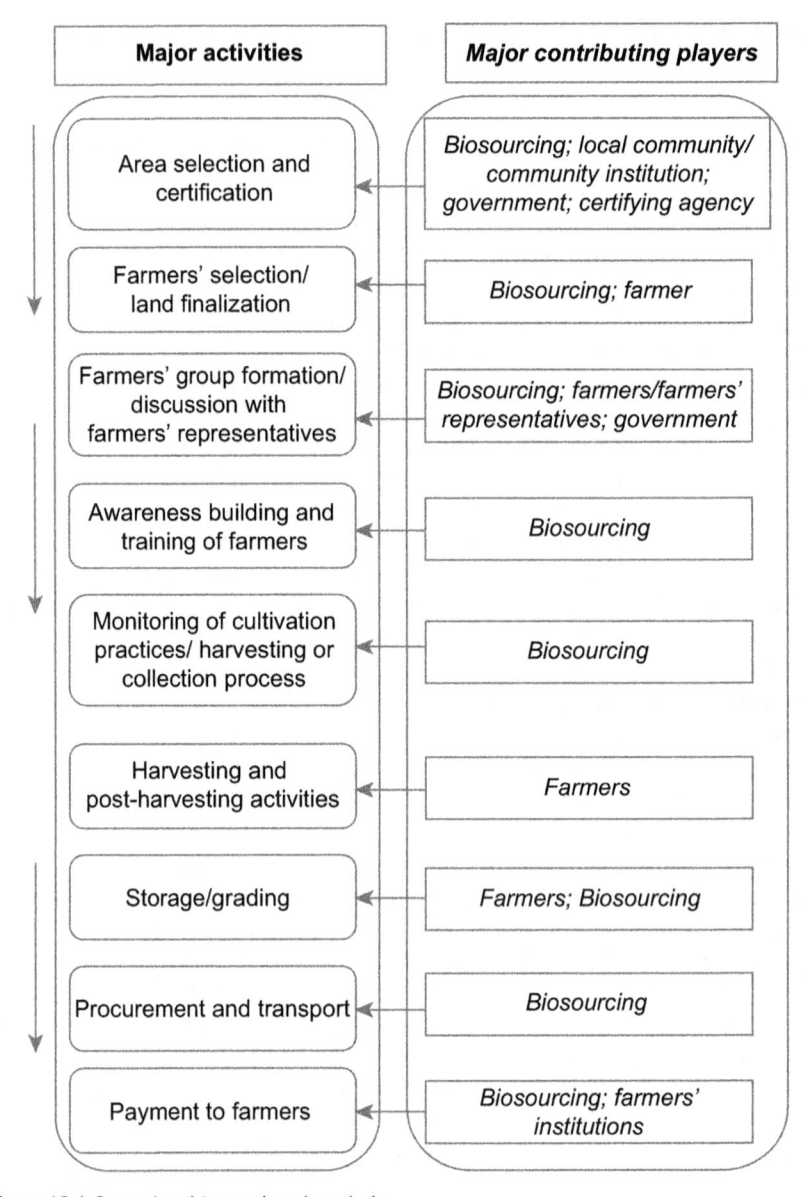

Major activities	Major contributing players
Area selection and certification	Biosourcing; local community/ community institution; government; certifying agency
Farmers' selection/ land finalization	Biosourcing; farmer
Farmers' group formation/ discussion with farmers' representatives	Biosourcing; farmers/farmers' representatives; government
Awareness building and training of farmers	Biosourcing
Monitoring of cultivation practices/ harvesting or collection process	Biosourcing
Harvesting and post-harvesting activities	Farmers
Storage/grading	Farmers; Biosourcing
Procurement and transport	Biosourcing
Payment to farmers	Biosourcing; farmers' institutions

Figure 16.1 Operational turmeric value chain

corrective measures. However, once the product has been cleared from a farm the company neither returns the product to farmers nor reduces the agreed price, irrespective of the result of the laboratory test. The finished product sample is again sent to the laboratory for final testing and the Certificate of Analysis (COA) is generated. All approved final products go to the organic warehouse. The COA is sent to the customer along with the product.

ORGANIC TURMERIC FROM EASTERN INDIA   167

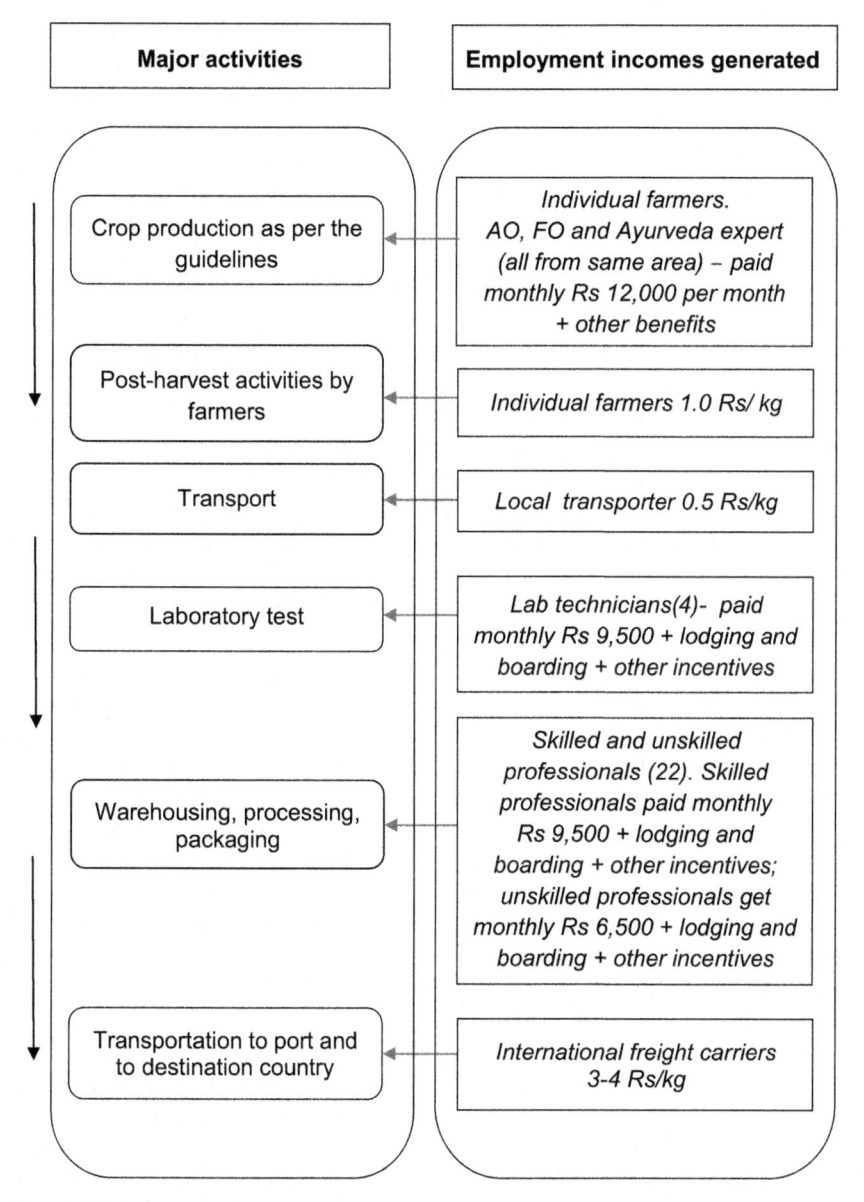

Major activities	Employment incomes generated
Crop production as per the guidelines	Individual farmers. AO, FO and Ayurveda expert (all from same area) – paid monthly Rs 12,000 per month + other benefits
Post-harvest activities by farmers	Individual farmers 1.0 Rs/ kg
Transport	Local transporter 0.5 Rs/kg
Laboratory test	Lab technicians(4)- paid monthly Rs 9,500 + lodging and boarding + other incentives
Warehousing, processing, packaging	Skilled and unskilled professionals (22). Skilled professionals paid monthly Rs 9,500 + lodging and boarding + other incentives; unskilled professionals get monthly Rs 6,500 + lodging and boarding + other incentives
Transportation to port and to destination country	International freight carriers 3-4 Rs/kg

Figure 16.2 The economics of the turmeric value chain

The final product is processed and packed as per the request of customers. If it has to be supplied in bulk a pack of 25 kilos is prepared, and for retail sales packaging is usually in 80 gram pouches. A sample of the packaged product along with the lot and order number is kept by the company for future reference.

Biosourcing mainly focuses on international markets for their products and depends heavily on online marketing. They have developed 75 websites on various organic herbs and spices and spend large sums to ensure that these websites come up first on any search engine for the related keywords. The URL for some of these websites, for example, are: organic.co.in, www.aloe-vera.co.in, www.amlaoil.com, www.basilleaves.com, www.buyginger.com, www.garlic-source.com, www.neemproduct.com, www.turmericpowder.net, www.pure-honey.net and many others.

Biosourcing receives enquiries through its websites and via emails which are forwarded to product managers who evaluate each enquiry and buyer. If the enquiry is considered to be genuine, the company checks the quantity, quality and price at which they can offer a product to the potential buyer. An offer letter is sent to the buyer, which is followed up by email and phone calls. Price negotiation also takes place at this stage. Once a price is agreed, a sample along with the COA is sent to the customer for the final order. Once the sample is approved by the buyer the order is closed. In most cases partial payment is received as an advance and final payment is made after delivery. The company keeps the sample along with the order number for reference, usually for one year. The product is packed and sent to the buyer by sea, mostly through Kolkata.

There are two different kinds of buyers. The first are bulk consumers such as hotels and restaurants, or trading companies that supply to local companies in their respective countries. The second category is those who sell the product as retailers using their own private labels. They request Biosourcing to package and label their orders in India before the product is shipped. For example, Pure Food Essentials, based in Australia, buys packaged products for retail sale in 80 gram packs. These products are either sold online, for example at www.purefoodessentials.com, or in local retail outlets. Although the exact cost of business for the importing companies, after the consignment is received at the destination port, could not be established for the purposes of this case study, it was found that the importing companies sold the products at margins varying from 50–200 per cent. For example, the retail online price of the 80 gram pack in Australia is US$5.95 or Rs298 rupees.

Biosourcing has been able to make its foreign buyers appreciate that India is a good source of organic products, that organic turmeric from Kandhamal is of high quality and that the company's entire business is transparent, efficient and professional. The customers are free to visit farms and physically verify the cultivation practices and conditions in which the crop is being grown, and many of them do this.

Benefits for the farmers

The erosion of tribal culture and traditional land-ownership, continuous alienation from the mainstream social system, and continuing neglect and harassment by government, have ensured that tribals continue to be poor, powerless and dispossessed. Biosourcing's intervention, however, is a welcome move that has made the tribals of Kandhamal both socially and economically more empowered. Biosourcing's intervention has resulted in the convergence of the efforts of different institutions and has in turn reduced duplication and costs. It has given more credibility to the partnering institutions, and has accelerated the pace of change.

At the field level, Biosourcing shows how village-level community institutions can be engaged for commercial transactions, and can thus create a win–win situation both for the community and for the company. In the last few years, the regular engagement of farmers and their institutions in the cultivation and marketing of turmeric has not only helped to strengthen local community-based institutions but has also increased the communities' confidence in their own institutions and traditional agriculture.

> **Box 16.1 Mr Naresh's story of turmeric and ginger cultivation**
>
> I have a family of six with my wife, two sons and two daughters. I and my sons were working in Vishakhapatnam as labourers. My wife stayed in Budurmila village with my daughters and worked as a daily labourer. We had a barren plot of half a hectare in our village. During the year 2005–06 we got to know about Biosourcing and contacted them. Their representative came to our village and encouraged us to use the land for cultivation of organic turmeric. For at least eight months two of their officers educated us about organic cultivation, how to use dry leaves, and how to make sheds of natural compost and how to use them. We were registered as organic farmers and then they started visiting us regularly. I started cultivating turmeric on 0.2 ha and on the rest I cultivated ginger. In 2007–08 I harvested 4 quintals of turmeric and 6 quintals of ginger. Biosourcing purchased my product giving Rs3 more than the market price, and since then I have stayed with Biosourcing. Today I have 5 hectares of land and I earn Rs220,000–250,000 per year. My daughters are married into good families and my elder son was married in December 2012. Today I live a happy life because of organic cultivation and Biosourcing.
>
> *Mr Naresh Pattamajhi of Budurmila, Tumudibandha*

Uncertified turmeric is currently sold in the market for Rs45 per kilo, whereas farmers who farm under the guidance of Biosourcing are getting Rs62 per kilo. Additionally, as Biosourcing encourages farmers to organically farm other crops too, their farmers are able to get a good price for all their crops. On average a farmer with 1 hectare of land who has partnered with Biosourcing for at least three years has been able to double his farm's income. In a few cases the total income of a household has gone up five times in the last six years, when the farmer has decided to grow only organic crops on his land.

The responses to the Progress out of Poverty Index (PPI) scorecard (developed by Grameen Foundation, available at www.progressoutofpoverty.org) by 30 randomly selected farmers revealed varying but positive results. Those farmers who had more than 4 hectares of land had reduced the likelihood of their being below the poverty line from 55.5 per cent to 42.1 per cent. Farmers with an average land-holding of between 0.5 and 4 hectares had also improved their position, but the likelihood of their being below the poverty line was still 65.2 per cent. Although the influence of other development interventions

170 COMMERCIAL AND INCLUSIVE VALUE CHAINS

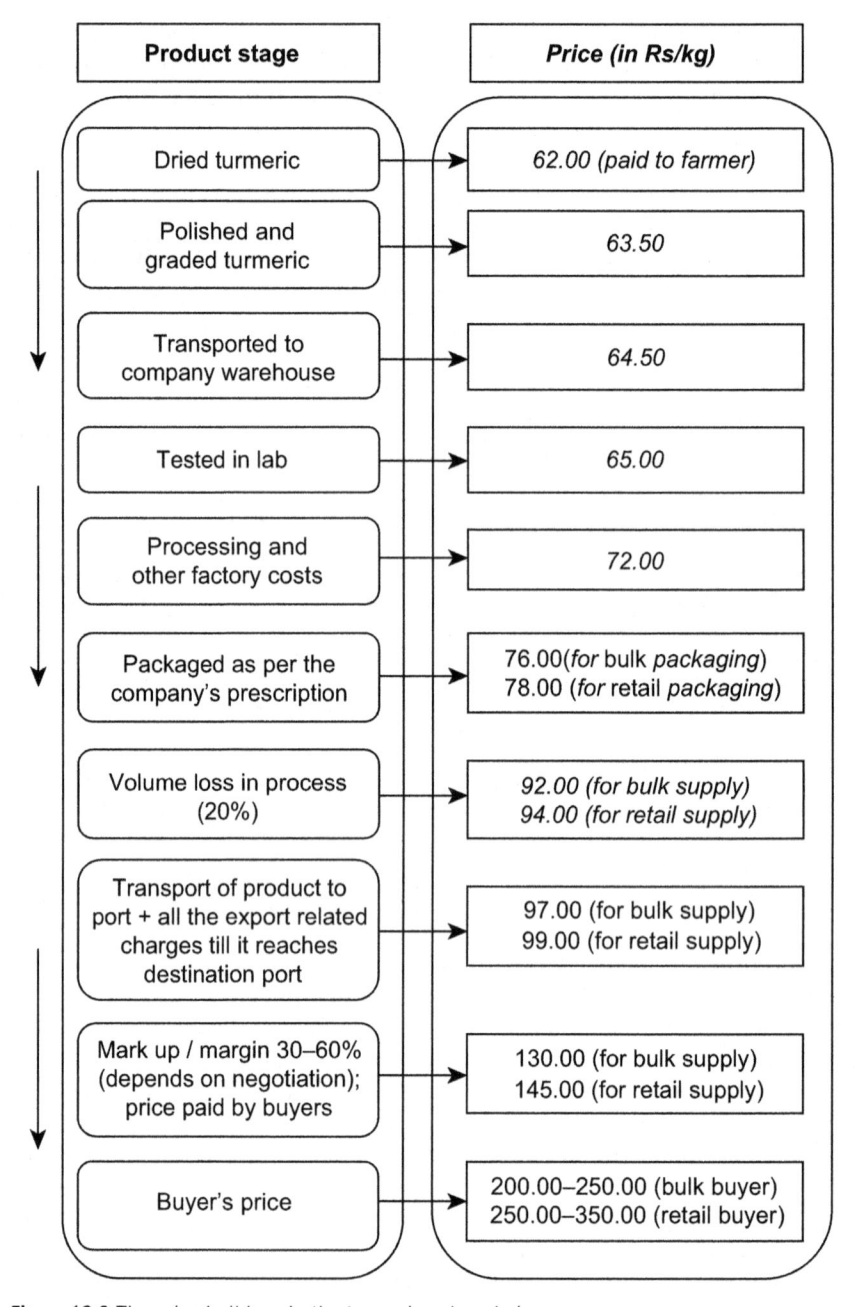

Product stage	Price (in Rs/kg)
Dried turmeric	62.00 (paid to farmer)
Polished and graded turmeric	63.50
Transported to company warehouse	64.50
Tested in lab	65.00
Processing and other factory costs	72.00
Packaged as per the company's prescription	76.00 (for bulk packaging) 78.00 (for retail packaging)
Volume loss in process (20%)	92.00 (for bulk supply) 94.00 (for retail supply)
Transport of product to port + all the export related charges till it reaches destination port	97.00 (for bulk supply) 99.00 (for retail supply)
Mark up / margin 30–60% (depends on negotiation); price paid by buyers	130.00 (for bulk supply) 145.00 (for retail supply)
Buyer's price	200.00–250.00 (bulk buyer) 250.00–350.00 (retail buyer)

Figure 16.3 The price build-up in the turmeric value chain

on the change in their PPI scores cannot be ruled out, the farmers agreed that their association with Biosourcing has substantially helped them to increase their incomes.

Conclusions

Biosourcing is still a young company, learning fast and determined to increase its reach on both sides: to the farmers from whom they source their products and to the customers to whom they sell them. The company has suffered many setbacks and still faces many challenges. It is, however, a 'green' business that not only takes care of the planet, the people and profit, but can also be replicated. In the turmeric value chain the farmers' share of the money paid by the consumers appears to be small, not because of inefficiency in the value chain, but because of the higher value addition that has made it possible for the product to travel from local to global. The farmers too have benefited. Although the nature of the business does not leave much scope for rapid growth, it does show how a well-managed value chain can benefit all the players. It requires joint efforts by farmers, government agencies and private institutions to make inclusive businesses like Biosourcing succeed, and to make organic farmers like those in Kandhamal more prosperous and more proud of their traditional culture.

References

Bhattacharya, P. and Chakraborty, G. (2005) 'Current state of organic farming in India', *Indian Journal of Fertilisers* 1(9): 111–123.

Charyulu, K. and Biswas, S. (2011) *Organic Input Production and Marketing in India*, Bangalore: Allied Publishers.

FAO (Food and Agriculture Organisation of the United Nations) (2004) *Green Outputs in India*, Rome: FAO.

New Indian Express (2014) 'Odisha wakes up to organic farming', *New Indian Express*, 23 November.

About the author

Dr Niraj Kumar is a member of the faculty of Rural Management at Xavier Institute of Management, Bhubaneswar, India. His areas of work are agribusiness, rural development and corporate social responsibility.

CHAPTER 17

Conclusions – what can we learn?

Malcolm Harper, John Belt and Rajeev Roy

Abstract

This chapter summarizes our findings across various value chains. It discusses the reasons for their start and their impact on individuals and communities. A SWOT analysis is done from the point of view of the lead firm in the chain. The overall objective is to present our learning in a consolidated manner that can be then used to drive the insights of practitioners and researchers interested in the field.

Keywords: value chains, inclusive value chains; SWOT analysis; lead firm; sustainability

Value chains in summary

A value chain approach can be defined as a set of interventions by chain actors, including buyers, processors, smallholders and other service providers to generate higher value and create win–win relationships among several chain actors. The value chains in this collection are no different, except that one or perhaps more than one set of participants happens to be made up of very poor people. In most cases, these poor people are the producers, but in some they are the customers, such as those who receive money through Dahabshiil, or retailers, such as the women who sell khat. The distinguishing feature of these value chains is that each of them assists many poor people to improve their livelihoods, often in a more sustainable way than through any development project, and at no cost to donors, non-government organizations or governments. Table 17.1 briefly summarizes some salient features of the 15 inclusive for-profit value chains.

There are many well-known and heavily publicized examples of inclusive value chains that have been set up by large multinational firms such as The Indian Tobacco Company, Unilever or Starbucks Coffee. It might have been expected that a book such as this would be made up mainly of value chains of this type, but we chose deliberately to seek out and study less well-known chains whose leaders are smaller businesses. Such companies are not interested in portraying themselves as 'socially responsible'; their aim is to build their businesses and to make money.

http://dx.doi.org/10.3362/9781780448671.017

Table 17.1 The case study value chains in summary

Product	Lead firm	Country	Approximate total of low-income producers or other households that benefit	Year of start
Turmeric and other herbal products	Biosourcing	India	1,500	2001
Banana Beer	Banana Investments	Tanzania	700	1989
Palm oil	Agro Selva	Peru	120	2009
Money Transfer	Dahabshiil	Somaliland	c. 4,200,000	1970
Granite	Various	India	?	?
Rice	Angkor Rice	Cambodia	50,000	1999
Cotton seed	Patidar Agro Centre	India	3,270	2001
Khat	–	Ethopia/Somaliland	?	?
Stove-liners and water filters	Chujio Ceramics	Kenya	360,000	1985
Millet flour	Nyirefami	Tanzania	400	1981
Green beans	–	Senegal	750	?
Cashew	–	India	5,500	?
Rice	Mtalimanja Holdings	Malawi	7,000	2012
Poultry	Suguna	India	15,000	1989
Milk	Moksha Yuga	India	15,000	2006

Lead firms

The term 'lead firm' refers to the firm that exerts a high degree of monitoring and control and defines the terms of trade among its network of suppliers, intermediaries or clients. Lead firms can provide strategic and organizational leadership in their network and can use their dominant position as the main innovator, coordinator of networks and standard-setter in order to maximize their own share of the profits which are made along the chain (Altenburg, 2006).

None of the chain 'leaders', with the possible exception of Dahabshiil, are multinational companies; they are local entrepreneurial businesses that have had to seek out and develop new suppliers and other business partners. More established and better-off producers, retailers and so on are generally already linked to existing value chains, and the companies in our cases have had to find new entities with which to collaborate. These may have operated 'below the radar' of older established firms, which often have a long history of transactions in their sector.

In this situation, newer smaller firms may find it difficult to participate in established supply chains. They have to look for new suppliers, disrupt existing procurement processes and reach deep into the supply chain and bypass existing powerful intermediaries. All this can be achieved by involving smallholders or other small-scale producers, intermediaries or customers who lack access to larger markets or suppliers. Small-scale entrepreneurs, smallholders, poor clients and other disadvantaged groups have limited resources and they lack substantial market presence. They tend to be ignored or exploited by very large suppliers or others, and are at a serious disadvantage in negotiations with them. Smaller firms, such as the leaders of the value chains described in these case studies, can often achieve a better match with the very smallest and weakest producers or others. They can enter into balanced business relationships with them and they are less likely to be held hostage to supplier power.

The businesses covered in this book have set up the value chains primarily driven by market forces and business logic. There was a natural fit between the businesses and the small producers, intermediaries or clients holders who make up the value chain.

Entrepreneurial value chains can help producers of traditional items to satisfy demands for more 'modern' versions of their products. For example, Nyrefami has been able to tap into customers' demands for packaged millet, and thus to enable smallholder millet farmers to preserve and to expand their sales, in spite of competition from other more sophisticated foodstuffs.

Small-scale local actors can also contribute their own understanding of local conditions that helps the lead firm. Dahabshiil is able to use local agents in both the remitting and the receiving communities to reinforce the requirement to know their customers. Similarly, and also in Somalia, the success of the khat value chain depends to a large extent on existing close family and clan relationships, and Agro Selva used abandoned oil palms in Peru, which larger firms had failed to use.

Larger producers can be wasteful and this can make many potential productive activities unviable, whereas smallholders can provide their full attention to a small operation. In Moksha Yuga, for instance, yields were improved dramatically but most of the dairy farmers have only a few cows. Effective micromanagement also helps maintain freshness in the khat value chain in Ethiopia, in a market where freshness of the product is highly valued.

Value chain initiatives led by profit-oriented entrepreneurs have some advantages over traditional development-oriented value chains promoted by NGOs and state actors. These traditional approaches have generally focused their efforts on the supply side and market conditions are not usually sufficiently incorporated in their models. Such assisted interventions may succeed as long as there is external support, but the entire effort may collapse as soon as the support ceases and market forces take over.

Benefits to small producers

Several of the value chains in this book involve provision of finance and other inputs to participating small units. Banana investments provided credit to aggregators of bananas, and Mtalimanja Holdings provides land, inputs and mechanized farming to the farmers who are under contract to the company.

Traditional livelihood sources often fail to bring in enough earnings for a family to subsist, but they are also very risky so that small-scale producers have to find alternative supplementary activities. Participation in a 'modern' value chain can reduce risk, so that smallholders no longer have to migrate to ensure that they earn enough, but in other cases, the new livelihood options are taken on as a supplementary source of revenue rather than as a replacement for older activities. In the case of the Suguna poultry value chain in India, for example, most of the people who have started a poultry farm also continue their traditional livelihood activities such as rice cultivation and vegetable farming. Suguna Company reduces the risks for its producers by providing proper feed and advice as well as veterinary medicines, and the price is assured. Traditional methods of moving international remittances involve a high degree of trust and may also break the law; Dahabshiil removes this risk, and is also able to offer the service at a lower price than competitors.

New production practices create higher value in the chain and improve efficiency. The lead firm may introduce such practices, and benefits from them, but the benefit is shared by the smaller participants in the chain. Biosourcing monitors the cultivation and collection of turmeric in order to ensure quality and conformity to organic standards, and Angkor Rice has introduced new varietals. The benefits of these improvements are not confined to the members of the value chain, and more successful smallholders and others may in time 'break out' of the value chain and start to build their own channels, as some of the Rajasthani cotton seed producers are doing. The chain leader has to replace these sources of supply, but the overall impact is highly positive.

One obvious difficulty of dealing with large numbers of small suppliers, or any other value chain participants, is the cost and management of large numbers of relationships with small-scale entities; individually they may be of small importance to the success of the value chain as a whole. The usual solution to this problem is to organize the smaller entities into groups, cooperatives or some similar institution that can speak for and to all its members and may also manage the logistics of collections, deliveries, payments and so on. This is of course a favourite strategy for governments and NGOs, which may in any case be politically or ideologically opposed to the emergence of private for-profit firms set up by individuals from a mass of partners to act as intermediaries or 'middlemen' (or women). This has been done by several of the chain leaders in our collection. Biosourcing has effectively made use of tribal groups which have already been promoted by government for 'joint forest management'.

It is significant, however, that in most of our cases the chain leaders make no use of groups of any kind; they prefer to deal with individuals. This may in part be because cooperatives have a bad reputation in so many countries, often because they have been 'used' by bureaucratic and political interests for a variety of purposes that do not always benefit the members, or as channels for subsidies and grants, thus destroying any sense of business economics. Or they have been promoted by NGOs whose staff have little appreciation of the realities of business and may, unconsciously or otherwise, wish to perpetuate the groups' dependence on themselves.

Groups are also difficult to manage, and their members often resent paying the 'market rate' for their managers, which may be many times more than their own incomes. Some groups are fortunate enough to find competent, honest and committed managers from among their own members, but this is often not possible; the poorer and more marginalized the membership, the less likely there is to be a suitable manager among them. These problems can be overcome, as Angkor Rice and others have found, and groups are more likely to succeed if they are not asked to do too much; it is hard for any group-managed entity to make fast decisions and to act in an entrepreneurial way.

Sustainable value chains tend to move away from one-off spot transactions toward long-term relationships, often between producers and processors or processors and exporters. The chain actors recognize that there is value in long-term associations above price taking. Agro Selva provides access to production facilities for a fee and also joint marketing of products. In the case of Angkor Rice, many formerly contracted farmers are now supplying directly to the market.

The value chains all contribute to local communities by creating sustainable local employment opportunities. Sometimes this is the major value that is added. The Senegalese green bean export companies, and most of the farmers themselves, prefer to be employed on larger company-managed farms rather than to be self-employed. Here again, this demonstrates a major difference between donor-promoted value chains and those that are described in this book; there is no nostalgic preference for independent microbusinesses, and if it makes business sense to provide jobs rather than to buy from traditional smallholders, that is what is done.

In the case of Suguna, not many of the independent poultry rearers employ anyone other than their own family members, but many of the farmers have stopped migrating to the cities for work, because rearing poultry for Suguna provides a better livelihood. The khat export value chain in Ethiopia and Somaliland generates many forms of indirect employment in transport, trading and other services. Khat is mainly retailed by small independent vendors who are located wherever there is a demand for their product; most of these are women, and khat vending is one of the most important sources of livelihood for women who may have no other means of support.

The granite value chain in Odisha has created large numbers of jobs, in the quarries and the cutting and polishing units, but many of these have been destroyed by ill-considered local government regulations. As so often happens, official efforts to 'do good' end up in doing a great deal of harm.

Smallholders, and the people who make a living as vendors or market traders, are often among the poorest in any society, and they are often the 'target beneficiaries' of development projects; NGOs, governments and others attempt to address their weaknesses by policy changes, or with subsidized credit, extension, training or other forms of assistance. The 'experts' who design these projects focus on the weaknesses of the people they are trying to assist; they sometimes achieve positive results.

The lead firms of the value chains in these case studies, however, did the opposite; they identified the strengths of the small units with which they are collaborating and effectively exploited these strengths, in the positive sense of the word, in order to achieve their objectives and maximize their profits.

'SWOT' analysis

Table 17.2 identifies some of these strengths, which can mean that such small informal units are the ideal partners for entrepreneurial firms that are trying to build new value chains in emerging markets where the more obvious partners are already fully engaged in existing networks, or where more formal or established potential partners do not yet exist or have withdrawn because of difficult conditions.

These people suffer from serious disadvantages, and are threatened by many modern trends, but the value chains described in this book have developed and prospered because their leaders have identified and are using

Table 17.2 Small poor players in value chains: 'SWOT' analysis from the point of view of value chain leaders

Strengths	Weaknesses
Low cost	Socially and physically 'remote'
Family labour	Illiteracy
Traditional knowledge and skills	Lack of language skills
Flexibility	Ill-health
Close social networks	Low aspirations and self-esteem
Less contact with competitors	Digitally divided from mainstream
	'Weaker sections'

Opportunities	Threats
Low-input organic farming	Standardization
Custom-designed individual products	Bulk requirements
Fair trade preferences	Inexorable growing worldwide inequity
Resurgence of cooperatives	Government subsidies and preferences

the strengths of the small-scale players, rather than trying to remedy their weaknesses.

There are many features that explain the success of the various value chains, and there is certainly no common model to which they all conform; one of the strengths of 'natural' value chains of this kind is that they are built on the basis of local circumstances, not according to any particular 'log-frame' or other development fashion.

There are some features, however, that are worth noting, even though they are not true of all the 15 value chains. The small-scale 'micro-participants' are generally undertaking familiar activities, or small modifications of such activities, that do not require training in new practices. This makes it easier for them to participate, and also minimizes the training and development costs of the lead firm.

Many of the participants have access to alternative markets of comparable value; this minimizes the risk of monopoly and exploitation, and the assets that they have to use for their participation in the value chain are generally not specific to that particular business. They can redirect their efforts if they wish or have to leave the value chain. They can also if necessary access alternative sources of raw material and supplies; this constrains any tendency towards monopolistic behaviour by the chain leader.

There are also a number of potential and actual problems that can prevent an inclusive value chain from fulfilling either its economic or its developmental potential. As is so often the case, the very poorest people, with no assets or skills, are usually excluded from direct participation in value chains. Inclusive value chains, like microfinance and so many other development initiatives, are not a solution for extreme poverty.

If left unchecked, the lead firms may start to exploit their micro-partners. The easiest way to increase short-term profits is to squeeze the small producers. This cannot be a long-term strategy, because the success of any value chain depends on the success of all its members, but without safeguards, there can be cases of contractual defaults, intimidation and unfair contracts. And, as a last resort, businesses may default or walk out of the chain if it becomes unmanageable or cheaper sources of supply become available.

The small-scale participants may also fail to deliver what is required of them. They may fail to comply with exacting standards because of lack of skills or finance; this can very easily happen if the value chains are badly designed or poorly resourced.

The lead firms in our case studies generally have the 'first mover advantage', but unless they are willing and able to make long-term investments in the value chain they may lose this to stronger competitors who can move in and disrupt the market by offering temporary high prices and other inducements designed to lure away producers.

Each case study includes some attempt to assess the impact of the value chain, either with the Progress out of Poverty Index (PPI) tool or otherwise. The samples are usually small, and may not be representative, and it is of course not possible to attribute changes in household well-being to membership of the value chain with 100 per cent certainty. We feel reasonably confident, however, that the small-scale producers, the employees, the vendors and the clients in the value chains are benefiting significantly from being part of them, and that these benefits are at least as sustainable as those that they might enjoy if they were participating in a government- or donor-assisted value chain.

We do not, however, suggest that the market can do everything, or that laissez faire policies will in time enable even the poorest producers to achieve decent livelihoods through membership of inclusive value chains. In spite of many positive experiences, such as those that are described in this book, the overall global trends are clear, and are not encouraging. Growth is often linked to increasing inequality, high growth does not automatically lead to decreases in the numbers of poor people or to improvements in the living standards of the poor, particularly of children.

The cases in this collection are not typical of all inclusive value chains, and even these value chains may in future evolve in different and less equitable directions. Smallholders and other poor participants may drop out of value chains; poor people have less assets, less skills, less access to information networks, less ability to take risks, less ability to adapt to changes and less power to change unfavourable situations.

Adam Smith's 'invisible hand' of the market will not ensure that poor people will automatically be included. Markets are amoral, they are merely a mechanism to distribute resources in society, and the private sector does not automatically include the poor.

The role of NGOs, governments, and donors

So what can NGOs, governments, and donors do to encourage private for-profit businesses to build and sustain inclusive value chains? Direct subsidies are not likely to be the answer; they tend either to promote value chains that are fundamentally unviable without subsidy, and that therefore do not survive its withdrawal, or to go to businesses that would have built the same value chains without the subsidy.

There is a basic mismatch between the donor world, with its focus on transparency, on planning, accountability, on paperwork and on measurable social impact, and the business world, which focuses on getting things done, on seizing opportunities and on profit, and this divide is not easy to bridge.

It may be more productive for donors and governments, and NGOs, to concentrate on more obviously 'social' institutions, on education, health, policies and the rule of law. In many poorer countries, these basic functions are woefully inadequate, and they should surely not be neglected in favour of much more difficult attempts to partner with private businesses.

If commercial enterprises and their value chains are to be assisted, it may be more productive to support them with the instrument they are used to, which is investment, not grants. Investing, through venture capital or soft (but still serious) loans, requires the investee businesses to ensure that their business models will generate the necessary repayments, and repaid investments can of course be reused to support other value chains.

Donors and NGOs should also employ people with business skills and experience and should, like the business managers in our case studies, aim to build on the strengths of the poor, rather than directly addressing their weaknesses. They should also learn from inclusive value chains that work without subsidies, such as those that are described in this book.

We are not of course against subsidies as such, and we have to acknowledge that most value chains, anywhere, and particularly those that involve farm products, are heavily subsidized. We hope that these case studies will show however that subsidies are not always necessary, and that they will provide some clues as to how subsidies can be used more effectively.

Reference

Altenburg, T. (2006) 'Donor approaches to supporting pro-poor value chains', Report prepared for DCED Working Group on Linkages and Value Chains, Amsterdam: DCED.

About the authors

Malcolm Harper is Emeritus Professor of Cranfield University in England, and has taught at the University of Nairobi and Cranfield University; he has worked since 1970 in enterprise development, microfinance and other approaches to poverty alleviation, mainly in India.

John Belt works for the Royal Tropical Institute (KIT) in Amsterdam. He is an agricultural economist with more than 20 years' field experience in agricultural development.

Rajeev Roy was an entrepreneur and currently teaches entrepreneurship. He has been engaged in several entrepreneurship development initiatives across the world. He is mentoring several start-ups in India.

Index

Action Aid 149–50
Agro Selva *see under* palm oil, Peru
Ahmed, Salehuddin 56
Angkor Rice, Cambodia *see under* rice
animal feed 26, 155, 158
Anti-Slavery International 62

banana beer (BIL), Tanzania
   business 27–30
   incomes of suppliers and employees 34–5
   markets and sources *29*
   processing 31–2
   production 30–1
   retail and consumption 32
   secondary actors 33
   taxes 32, 35
   trade 31
   value chain 32–3, *34*
   wholesale and distribution 32
Barclays Bank, UK 72, 77
Biosourcing *see under* turmeric, Odisha
bonded labour 61–2

Cambodia *see under* rice
cashew nuts (Trimurti Cashew Industries), Odisha 143–4
   agents as aggregators 146
   pressure from international development community 149–50
   prices 147–8
   processing units 146–7
      employment conditions 148–50
   producers 145–6
   supply chain 144–8
   traders 147

child labour 38, 45
   and bonded labour 61–2
childcare facilities 149
children's education 40, 72, 87, 112, 113, 149
China 64, 90, 93, 98
Chujio Ceramics *see* stove liners (KCJ/Chujio Ceramics), Kenya
community councils 134
community-based institutions 163–4, 168–9
contract farming 98–9, 133
cooperatives
   disadvantages of 176–7
   failure of Operation Flood, India 105–6
cotton seed, South Rajasthan 37–40
   agents 40–1, 43–5
   farmers 37, 39
      income 40–1
      poverty reduction 41–2
   ginneries 39, 43, 44, 45
   and Gujarat 38–9, 42, 45
   international seed companies 39
   organizers 39, 40–1
      Patidar Agro Center Company 42–3
   value chain *40*
      benefits accrued from 40–1
credit/loans
   Banana Investment Ltd (BIL) employees 35
   crop production 44–5, 82–3, 94, 143
   poultry production 118, 120
   stove liner production 51

Dahabshiil *see under* remittance companies, Somalia/Somaliland
dairy, India
　development (Operation Flood cooperative model) 105–6
　Karnataka (KMF) 106–7, 109, 110
　Mosha Yug Access (MYA) 107
　　business model 108–9
　　case studies 111–13
　　challenges 107–8, 111
　　farmer benefits 110–11
　　rural infrastructure 109–10
　　value chain *113*
decentralized model 117–18
donors and NGOs 1, 4, 173, 180–1
Duale, Abdirashid 75
Duale, Mohamed Saeed 75

entrepreneurial value chains 174–5
environmental impacts
　illegal granite quarrying 61
　palm oil production 152
　stove liners 47–8, 55–6
EU requirements/standards 130, 131–3, 165
exports *see* cashew nuts, Odisha; granite, Odisha; green beans, Senegal; khat, Ethiopia and Somaliland; turmeric, Odisha

Forest Protection Committees (VSS) 164
fresh fruit and vegetable (FFV) exports 127–8, 134

GlobalGAP 130, 131, 132
Grameen Foundation 6
granite, Odisha
　cutting and polishing 66–7
　　employees 66, 67
　and India
　　regulatory problems 60–2
　　worldwide industry 59–60
　quarry 63–6
　　employees 64–6

　extraction process 64
　licensing 60, 61, 64–5
　value chain 62–3, 68
　wholesalers 68
green beans, Senegal 129–30
　certification schemes 130, 132
　export supply chain 130–1
　　agro-industrial production and employment 134
　　companies and consolidation 131–2
　　rural households 134–6, *137*
　　smallholder contracts *vs* large-scale production 133
　　vertical coordination and ownership integration 132–4
　fresh fruit and vegetable (FFV) exports 127–8, 134
　income effects 136–40
　poverty reduction effects 140–1
　regulations/standards 130, 131–3
groups, disadvantages of 176–7

Harper, M. 22, 25
Human Development Index (UNDP) 37

incomes
　banana beer 34–5
　cotton seed 40–1
　green beans 136–40
　khat 21, 23, 24, 25
　palm oil 156–7
　poultry farming *123*
　remittances 72
　see also poverty reduction
Indian Committee of the Netherlands 62
information technology
　mobile phones 74, 78
　online marketing 166–8
International Crops Research Institute (ICRISAT-HOPE) 85

Kenya Ceramic Jiko (KCJ) *see* stove liners, Kenya
khat, Ethiopia and Somaliland 17
   advantages and disadvantages 20–1, 25–6
   cultivation 17–18
   economic impacts 20
   farmers 20–1
   Hargeisa market 22
   origins of use 18
   post-harvest operations 21
   prices and incomes 25
      'caretakers' 24
      drivers 24
      farmers 21
      vendors and assistants 23, 24
   taxes 21, 22, 23
   traders 20–1, 24
   transport 18–19, 21, 22, 24–5
   types and consumers 19–20, 21, 25
   value-added at each stage 24–5
   vendors 22, 23–4
Kirk Natural Stone, Scotland 62

large-scale producers
   palm oil plantations (OLAMSA), Peru 151–2, 156
   and small producers, comparison between 175
   see also green beans, Senegal
lead firms 174–5
loans *see* credit/loans

Maimbo, S. 72
Malawi *see under* rice
millet (Nyirefami Ltd), Tanzania
   business 81–3
   costs, profits and margins 85
   markets and sources *86*
   poverty reduction effects 86–7
   processing 84
   production 83–4
   retail and consumption 84–5
   secondary actors 85
   traders
      and agents 84
      testimony 87
   value chain 83–6
   wholesale and distribution 84
mobile phones 74, 78
Moily, Harsha 107
Mosha Yug Access (MYA) *see under* dairy, India
Mtalimanja Holding Ltd (MHL) *see under* rice

National Rural Employment Act (NEGRA), India 64–5, 66
NGOs and donors 1, 4, 173, 180–1
Nyirefami Ltd *see* millet, Tanzania

Odisha *see* cashew nuts; granite; turmeric
organic farming *see* turmeric
ownership integration, vertical coordination and 132–4

palm oil, Peru
   Agro Selva 152–3
      end market and production costs 155–6
      future of 158–9
      income 156–7
      poverty reduction 157–8
      value chain 153–5
   large-scale approach (OLAMSA) 151–2, 156
Patidar Agro Center Company 42–3
Potters for Peace (PfP) 52–3
poultry, India
   industry 115–16
   PRADAN approach 118–19
   Suguna 116–18
      case study 123
      changes among smallholders 119–22
      decentralized model 117–18

employees 122, *123*
livelihood portfolios 121–2
survey 119
poultry and pig feed, Peru 155, 158
poverty reduction/PPI 6–7, 179–80
    banana beer production 34–5
    cashew nut processors 149
    cotton seed cultivation 41–2
    dairy farmers, case studies 111–13
    granite workers and processors 65, 67
    green bean exports 140–1
    millet farmers 86–7
    palm oil farmers 157–8
    poultry farmers
        case study 123
        and employees 122, *123*
    remittance recipients 77
    rice farmers 94–5, 100, *101*
        case studies 103–4
    turmeric farmers 169
    see also incomes
PRADAN, India 118–19
Progress out of Poverty Index (PPI) *see* poverty reduction/PPI

Rajasthan *see* cotton seed, South Rajasthan
remittance companies, Somalia/Somaliland
    advantages of 71, 72, 75–6
    agents 74, 76–7
    and Barclays Bank, UK 72, 77
    civil war and migration 72, 75
    Dahabshiil 71, 72, 74, 75
        customers and recipients 76–8
        future of 78
    money flow map *73*
    transfer process and costs 73–4, 76
    value chain *73*
rice
    Angkor Rice, Cambodia
        background 97–8
        case studies 103–4
        contract farming 98–9
        criticisms 102–3
        farmer benefits 100–1
        future of 103
        un-contracted farmers 102
        value chain 99, *100*
    Mtalimanja Holding Ltd (MHL), Malawi 90–2
        background 89–90
        farmers and other actors 92–4
        impact on poverty 94–5
        input suppliers 92, 94

self-help groups (SHGs) 164
Selian Agricultural Research Institute (SARI) 30–1, 33
Senegal *see* green beans, Senegal
small producers
    benefits to 176–8
    and large producers, comparison between 175
    'SWOT' analysis 178–81
Somalia/Somaliland *see* khat; remittance companies
stove liners (KCJ/Chujio Ceramics), Kenya 47–8, 57
    early lessons learned 51–2
    impact on households and environment 47–8, 55–6
    marketing and supply chain 50–1
    start-up 48
    technology
        as competitive advantage 49–50
        development and expansion 52–3
    value chains 53–5
    and water filter production 52–3, 54–5
subsidies 4, 180
Suguna *see under* poultry, India
sustainability 4, 177
Swedish International Development Cooperation Agency (SIDA) 33
'SWOT' analysis 178–81

Tanzania *see* banana beer (BIL); millet (Nyirefami Ltd)
technology
 agricultural 92–3, 152–3
 ceramics 49–50, 52–3
 mobile phones 74, 78
 online marketing 166–8
Trimurti Cashew Industries *see* cashew nuts, Odisha
turmeric, Odisha
 Biosourcing 161, 162–3
  buyers 168
  certification and standards 163, 165
  community-based institutions 163–4, 168–9
  cultivation and post-harvest operations 164–6
  online marketing 166–8
  poverty reduction 169
 government role 164
 organic cultivation in India 161–2
 tribal people 162, 163, 168
 value chain *166, 167*

United States African Development Foundation (USADF) 82
University of Dar es Salaam (UDSM) 33

value chains
 approach 3–5, 173
 lead firms 174–5
 see also small producers
vertical coordination and ownership integration 132–4

Wambugu, Kamwana *see* stove liners (KCJ/Chujio Ceramics), Kenya
water filter production 52–3, 54–5
women's employment 118–19, 139, 148–50
working conditions 61–2, 148–50

About CTA

The Technical Centre for Agricultural and Rural Cooperation (CTA) is a joint international institution of the African, Caribbean and Pacific (ACP) Group of States and the European Union (EU). Its mission is to advance food and nutritional security, increase prosperity and encourage sound natural resource management in ACP countries. It provides access to information and knowledge, facilitates policy dialogue and strengthens the capacity of agricultural and rural development institutions and communities.

CTA operates under the framework of the Cotonou Agreement and is funded by the EU.

For more information on CTA, visit www.cta.int

About KIT

KIT is a not-for-profit knowledge and expertise organization for international and intercultural cooperation. KIT's aim is to provide innovative and practical solutions primarily for low and middle-income countries. Our work encompasses the areas of research, advice, training, and education for international organizations, companies, NGOs, governments, and students, with specific emphasis on improving healthcare and facilitating sustainable economic, social, and organizational change. For more information visit http://www.kit.nl/kit/en/